MINING AND THE FRESHWATER ENVIRONMENT

MINING AND THE FRESHWATER ENVIRONMENT

MARTYN KELLY

Department of Botany, University of Durham, UK

with contributions by

W.J. ALLISON, A.R. GARMAN and C.J. SYMON

formerly Chelsea College, University of London, UK

ELSEVIER APPLIED SCIENCE
LONDON and NEW YORK

ELSEVIER SCIENCE PUBLISHERS LTD
Crown House, Linton Road, Barking, Essex IG11 8JU, England

Sole distributor in the USA and Canada
ELSEVIER SCIENCE PUBLISHING CO., INC.
655 Avenue of the Americas, New York, NY 10010, USA

WITH 24 TABLES AND 37 ILLUSTRATIONS

Softcover reprint of the hardcover 1st edition 1988

First Edition 1988
Reprinted 1991

British Library Cataloguing in Publication Data

Kelly, Martyn
Mining and the freshwater environment.
1. Freshwater ecosystems. Effects on mining
I. Title II. Alison, W.J. III. British Petroleum Company
574.5′2632

ISBN-13: 978-94-010-7105-5 e-ISBN: 978-94-009-1359-2
DOI: 10.1007/978-94-009-1359-2

Library of Congress CIP Data applied for

Typeset at The Alden Press, London, Northampton, Oxford

Foreword

Over the last century the discharge of crude or partially treated sewage has probably been the most widespread, most documented and certainly the best understood form of pollution entering the aquatic environment. In the past two decades, however, there has been an increasing public awareness of the potential hazards that exist from the contamination of the freshwater environment by toxic substances associated with the mining industry.

World demand for minerals has intensified the exploitation of natural resources. In most western and newly developed countries significant mining proposals are now strenuously regulated to protect the environment. These involve economic and legislative measures and the use of appropriate control technologies. This concern will undoubtedly continue to spread worldwide requiring a programme of enlightened environmental protection management policies and practices for the future.

This book has been prepared as a synthesis of our current understanding of the effects of various heavy metals and acidic discharges likely to contaminate the freshwater environment as a direct result of mining activities. The review is based upon the dissertations of former BP sponsored students who were engaged to provide a better scientific understanding of the causes of environmental problems associated with this industry.

It gives us great pleasure to publish this information for use and application by a wider audience as part of the contribution of The British Petroleum Company p.l.c. to European Year of the Environment.

DR CHRISTOPHER GIRTON
BP International Limited

Preface

There is a popular misconception that it is impossible for industry and nature to live happily side by side; the requirements of each are presumed to be mutally exclusive. As with so many aspects of modern life, the truth is that with forethought, planning and, it must be admitted, capital expenditure on the part of the industrialist, nature can exist quite happily alongside an industrial plant. Perhaps the pendulum has swung too rapidly from the 'dark, satanic mills' of Blake's Victorian England to the modern environmentally-conscious and much talked-of 'green vote'. Mrs Gro Harlem Brundtland, Prime Minister of Norway, when talking to industrialists, said 'You . . . are often regarded as the prime source of environmental problems, especially pollution; however, you are also seen as the source of many comforts, convenient machines, clothes, cars and many basic ingredients of a good life. You are also the source of development and of jobs. You are thus both reviled and appreciated, very often by the same people, although at different times.' (quoted by Elkington, 1987).

That is not to say, of course, that the campaigning of these environmentalists has not won many valuable campaigns — Rachel Carson's 'Silent Spring' (1962) and the subsequent outcry over irresponsible use of pesticides is probably the best known example — but the ball is now back in the court of the industrialists and the environment is firmly on the mainstream political agenda. One example of this, 'European Year of the Environment' declared by the member states of the European Community for March 1987 to March 1988 had the stated aim of increasing environmental awareness, and promoting further incorporation of environmental considerations in the formulation of economic, industrial and agricultural policies.

The technology now exists to keep the developed world in the state to which it has fast become accustomed, but at the same time to do this responsibly and with the longer-term interests of the environment very much to the fore. With the rising costs of fuels and raw materials it is in the interest of industry to develop processes that create less waste. Investment in the resources associated with environmental protection, in turn smooths the path for industrial development by anticipating the legitimate concerns of local naturalists about proposed developments at an early stage in the planning.

There is more than a grain of truth buried within the old saying 'the solution to pollution is dilution'. Natural transport and transformation processes may spread a small amount of toxic material concentrated in a small area into a larger area, such that its concentration is no longer toxic. This does not mean that it is perfectly reasonable for a factory to periodically tip its effluents into the nearest river and forget about it. The extent to which waste can be transformed and the concentration at which it ceases to have a detectable effect vary enormously. Not only do the types of pollutant vary but their behaviour also depends upon the nature of the receiving environment. Under these circumstances, the advice of an environmental scientist is essential.

Mining is a particularly notorious industry from the point of view of pollution. It embodies a great many of the virtues outlined in the first paragraph; in fact few other industries can have had such a large hand in shaping the present economic status of the developed world. Extraction of coal and minerals, however, often involves the removal of tonnes of overburden to expose veins or seams, only a few percent of which may be of economic value. The problem then is what to do with the waste.

My introduction to this was in the Northern Pennine Orefield in the north of England. At the turn of the century, this area was the world's largest supplier of lead; however by the early part of this century, richer and more profitable supplies were found overseas and the area went into decline. The area is now predominantly agrarian, although from time to time entrepreneurs have tried to open up some of the old mines or have worked through the spoil heaps trying to extract more lead, during surges in its market price. More recently a company has found it worthwhile to mine fluorspar, once regarded as an unwanted 'gangue' mineral but now of value in its own right.

Despite these attempts, the landscape is still much as the miners left it, abandoned adits driven into hillsides, spoil heaps beside these, barely vegetated even after all this time and, here and there, the remains of smelt mills where ore was refined. Out of many of the adits flows water which, having percolated through the overlying Carboniferous Limestone, drains along the old levels over the lead and zinc-rich minerals and into the streams and rivers. During and after rainfall, water which falls on the spoil heaps dissolves some lead and zinc and carries more in suspension into the streams and rivers; during dry spells in the summer dust blows from these heaps onto adjacent farm land, occasionally poisoning cattle, or into streams and rivers. The miners' memorial may be thousands of church roofs, miles of lead pipes and the ammunition which defended an empire but their bitter legacy remains just where they left it. Heavy metals are not a problem that will just go away.

However, we can start with an understanding of how they behave in ecosystems; what chemical forms they adopt, which forms are accumulated by organisms, what concentrations are toxic, how they are passed along food chains and, eventually, what risk there is to man himself. Also, we can compare different sites and see if the answers to these questions are the same at each. If not, what other factors in the environment may cause these differences? From here, the environmental scientist can go on to suggest ways of treating effluents that reduce the input of metals. In the case of spoil heaps, revegetation not only reduces leaching of toxic materials but may also improve the visual and amenity value of the site. Furthermore, the knowledge gained from studies of existing sites allows predictions to be made about effects at future sites.

The primary literature, though enormous, is scattered through many journals. Many case studies especially appear only in regional publications and are thus lost to anyone who does not have access to a comprehensive library. A review of this kind, hopefully, circumvents many of these problems. It is not comprehensive — the field is now too large; however it does bring together a lot of literature and summarizes the main findings. Wherever possible, I have tried to do this using graphs and diagrams to show more clearly the differences between groups. Several parts were based upon three literature searches performed by students (W.J. Allison, A.R. Garman and C.J. Symon) on the Applied Hydrobiology MSc course at Chelsea College, University of London and are confined to the subjects of their study. These were four metals: nickel, copper, zinc and lead plus coal; however it was impossible to restrict discussion to these. The aim of the book is to draw out generalities so that part of what is written will apply to other metals and, equally, some relevant results using metals such as cadmium have been included to help make particular points. Placer gold mining was included as a case study partly for its own intrinsic interest and partly because it focusses attention on the highly important aspects of siltation and turbidity which are often overlooked in studies of the effects of mining.

No scientist works in a vacuum; in my case I owe a great deal to the guidance and advice given by Brian Whitton during my PhD studies at Durham University although many others, through conversation and criticism, have contributed. The execution of this project owes much to financial support and encouragement from Group Environmental Services, BP International Limited, especially Chris Girton. Comments on early drafts by Patrick Denny (Queen Mary College, University of London), Malcolm Hutton (Monitoring and Assessment Research Centre, University of London) and Tom Robinson (BP Coal) were invaluable whilst staff at Elsevier Applied Science Publishers eased the transition from manuscript to finished product. Finally, my wife, Heather, deserves mention not only for proof-reading my work but also for her forebearance and for tolerating the gradual encroachment of her home by piles of my papers.

MARTYN KELLY

Contents

Chapter 1

Geological and Economic Background

Environmental factors associated with mining, as distinct from those associated with the subsequent use of the minerals, are largely confined to relatively restricted regions in the vicinity of the appropriate geological formations and downstream in the catchment. Nonetheless, it is not the actual extraction of the mineral so much as the disturbance thus caused which produces the pollution; 50 or 100 years after the cessation of mining activities upland regions in Britain are still polluted by water draining out of disused mines and spoil heaps.

In other words, on a world scale, if you can identify those areas which contain the mineral then you know which areas are going to be subject to the effects of mining. In the modern age, with heavy metals used so widely, their effects are far more widespread; however the effects of mining *per se* remain highly localized. In Great Britain mining for metals was confined to upland regions with harder rocks, which, in turn, were less biologically productive than lowland regions. Coal mining, on the other hand, is found in upland and lowland areas.

1.1 NICKEL

There are two sources of nickel: sulphide ores which account for 60% of world output and laterite ores which account for the remainder. Nickel sulphide deposits are usually between 1 and 4% nickel and may also contain significant amounts of copper and platinum group metals; the ratio of nickel to copper is usually in the range 10:1 to 1:3. Laterite deposits contain about 1–3% nickel, but have only negligible amounts of copper. The principal nickel-bearing minerals of economic significance are shown in Table 1.1.

The current major producers are Canada (which produces over 50% of the free-world's supply from the Sudbury region of Ontario alone) and the USSR (Table 1.2). Despite a sharp increase in demand since the 1950s nickel production is still far below that of most other common metals. More than 75% of world production is used in the making of alloys; there are over 3000 different nickel-bearing alloys in use, of which stainless steel is the most important.

1.2 COPPER

The two main sources of copper are porphyry ores (44·8% of world output) and sedimentary, stratabound deposits (26·4%). The balance is supplied by deposits in metamorphic rocks, free (native) copper and 'supergene' deposits formed by the weathering of sulphide minerals (Wolfe, 1985). It has been estimated that porphyries will continue as the major copper source until the end of the century. There are about 150 different copper-bearing minerals of which chalcopyrite is the most abundant and widely mined (Table 1.1).

Currently the main producers are the USA and the USSR; other major producers include Canada, Chile, Zambia and Zaire (Table 1.2). Since copper was first mined there has been a trend towards exploitation of larger and lower grade copper deposits. In Europe in AD 1540, for example, copper ores consisting of 8% copper were mined. In 1890 this figure was 6% and in 1906, 2%. About half of the world output is used in the electrical and telecommunications field; most of the rest is used in the building industry and in general engineering fields.

1.3 ZINC AND LEAD

As these two metals are frequently associated in the field it is convenient to consider their geology together. The most common minerals are both sulphides: sphalerite (zinc sulphide, 'zinc blende') and galena (lead sulphide). Other common minerals are shown in Table 1.1.

Lead was apparently mined in Britain in Roman times (Raistrick & Jennings, 1965) although the greatest mining activity was during the period 1750–1900. During the second half of the 19th century the

Northern Pennine Orefield was the largest lead-mining region in the world until cheaper foreign imports led to its decline. In this region and in other lead-mining regions in Britain it was zinc, from sphalerite, present as a 'gangue' mineral, which was thought to be the major toxicant. Now the major producers of both zinc and lead are Australia, Canada, the USA and the USSR (Table 1.2).

Zinc is used largely in alloys, especially brass, and for galvanizing iron and steel. Other uses include paint and battery manufacture. Lead was used in the past for plumbing, for covering roofs and in bullets. Newer uses include manufacture of solder, paints, batteries and cable coverings. Tetra-alkyl lead was also widely used as an 'anti-knocking' agent in petrol although this is now being phased out in many countries.

1.4 COAL

Coal is the net result of two processes: first, an attenuation of the breakdown of organic remains of plants and animals, and second, the lithification of the accumulated materials. Attenuation of microbial breakdown commonly occurs under the anaerobic conditions encountered in wetlands, as the discovery in bogs of well-preserved human remains from the iron age testifies. The term 'coal' is a blanket term for a continuum from the barely to the fully lithified, depending upon the amount of compaction and heating it has undergone. From peat the series goes from lignite, sub-bituminous, bituminous and sub-anthracite to anthracite, a hard, shiny black coal with the highest proportion of carbon and the lowest proportion of other volatile constituents. One by-product of the anaerobic conditions is the reduction of sulphur to sulphide, most commonly as iron pyrites (Fe_2S). The sulphur content of coal commonly ranges fom 0·5–3·0%.

The rise of the Industrial Revolution in Britain was closely linked to the availability of coal as a source of energy and now, even after 200 years of exploitation, there are still significant reserves of coal in Britain. Other major producers include other members of the European Economic Community, Canada and the USA, the USSR and China (Table 1.3).

Table 1.1
Principal nickel, copper, zinc and lead minerals of economic significance. The most important mineral for each metal is shown in bold type

	Composition
Nickel mineral	
Annabergite	$Ni_3(AsO_4)_2, 2H_2O$
Bravoite	$(Ni, Fe, Co)S_2$
Gersdorffite	$NiAsS$
Mackinawite	$(Fe, Ni)_9S_8$
Maucherite	$Ni_{11}As_8$
Millerite	NiS
Niccolite	$NiAs$
Pararammelsbergite	$NiAs_2$
Pentalandite	**$(Ni, Fe)_9S_8$**
Skutterudite	$(Ni, Co)As_3$
Violarite	$(Ni, Fe)_3S_4$
(from Duke, 1980)	
Copper mineral	
Azurite	$2CuO_3, Cu(OH)_2$
Bornite	Cu_5FeS_4
Chalcocite	Cu_2S
Chalcopyrite	**$CuFeS_2$**
Covellite	CuS
Cuprite	Cu_2O
Enargite	$3Cu_2S, As_2S_5$
Malachite	$CuCO_3, Cu(OH)_2$
Native copper	Cu
Tetrahedrite	$3Cu_2S, Sb_2S_3$
(from Bowen & Gunatilaka, 1977)	
Zinc mineral	
Sphalerite (zinc blende)	**ZnS**
Zincite	ZnO
Franklinite	$(Fe, Zn, Mn)(Fe, Mn)_2O_2$
Smithsonite (dry-bone)	$ZnCO_3$
Hydrozincite	$Zn_5(OH)_6(CO_3)_2$
Willemite	Zn_2SiO_4
Hemimorphite (calamine)	$Zn_4(OH)_2Si_2O_7$
(From Cammarota, 1980)	
Lead mineral	
Lead	Pb
Minium	$2PbO.PbO_2$
Massicot	PbO
Plattnerite	PbO_2
Pyromorphite	$3Pb_3P_2O_8, PbCl_2$
Cotunnite	$PbCl_2$
Anglesite	$PbSO_4$
Cerusite	$PbCO_3$
Galena	**PbS**
(From Emmons, 1918)	

Table 1.2
World mine production of nickel, copper, zinc and lead in 1985, based on figures in *World Mineral Statistics* (1987)

Country	*Production (tonnes)*			
	Nickel	*Copper*	*Zinc*	*Lead*
European Economic Community				
France		223	40 600	2 500
Germany, West		900	117 600	20 500
Greece	10 900	300	21 107	19 752
Irish Republic			191 700	34 600
Italy		127	45 438	15 600
Portugal		2 563		
Spain		60 960	234 695	85 636
United Kingdom		592	5 043	3 994
Other European Countries				
Albania	10 000	15 000		
Austria			24 259	7 500
Bulgaria		78 000	68 000	97 000
Czechoslovakia		10 500	7 500	2 718
Finland	7 920	26 600	60 600	2 400
Germany, East	2 000	12 000[a]		
Greenland			70 300	17 800
Hungary			1 606	575
Norway	440[a]	18 969	27 783	3 349
Poland		432 000	190 900	51 300
Romania		30 000[a]	40 000[a]	35 000[a]
Soviet Union	175 000[a]	1 030 000[a]	1 000 000[a]	580 000[a]
Sweden		91 845	216 400	76 200
Turkey		28 700	37 400	10 000
Yugoslavia	7 500[a]	142 479	89 300	115 100
Africa				
Algeria		200	13 500	3 800
Botswana	19 560	21 703		
Congo		100		2 500
Morocco		32 800	14 662	106 784
Mozambique		124		
Namibia		47 600	31 200	48 600
Nigeria				300
South Africa	29 000	205 052	96 943	98 425
Tunisia			5 600	2 500
Zaire		502 100	67 925	
Zambia		519 600	31 900	15 062
Zimbabwe	9 897	20 700		

Table 1.2 — contd.

Country	*Production (tonnes)*			
	Nickel	*Copper*	*Zinc*	*Lead*
North America				
Canada	169 971	738 637	1 206 683	284 595
Cuba	33 400	3 100		
Dominican Republic	25 400			
Guatemala				
Honduras			44 000	21 300
Mexico		178 904	275 412	206 732
USA	5 558	1 105 758	226 545	413 955
South America				
Argentina		300	36 200	28 500
Bolivia		1 665	38 110	6 242
Brazil	20 300	50 000	123 811	18 200
Chile		1 356 400	22 288	2 473
Colombia	12 500			100[a]
Equador		100		200 600
Peru		384 600	582 600	
Asia				
Burma	20[a]	16 700	4 400	7 623
China	190 000[a]	190 000[a]	200 000[a]	175 000[a]
Cyprus		1 065		
India		48 000	53 000	26 000
Indonesia	48 200	88 750		
Iran		50 000[a]	50 000	21 600
Japan		43 208	253 021	49 951
Korea, North		10 000[a]	185 000[a]	80 000[a]
Korea, South		308	45 746	9 698
Malaysia		30 506		
Mongolia		136 000		
Oman		17 700		
Philippines	27 653	237 713	1 740	
Taiwan				
Thailand			68 300	19 654
Vietnam			10 000	
Australasia				
Australia	85 761	258 474	737 233	493 493
New Caledonia	72 372			
Papua New Guinea		175 048		
World total	797 000	8 500 000	6 900 000	3 500 000

[a] Estimated figure.

Table 1.3

World mine production of coal in 1985, based on figures in *World Mineral Statistics* (1987). No separation is made into different types of coal (anthracite, bituminous, brown coal and lignite)

Country	*Production (× 1000 tonnes)*
European Economic Community	
Belgium	6212
France	16984
Germany, West	203063
Greece	35962
Irish Republic	57
Italy	1892
Portugal	238
Spain	39663
United Kingdom	94047
Other European Countries	
Albania	2195[a]
Austria	3081
Bulgaria	30880
Czechoslovakia	28538
Germany, East	312156
Hungary	24042
Norway	570
Poland	249388
Romania	44500
Soviet Union	726000
Turkey	46296
Yugoslavia	69500
Africa	
Botswana	437
Morocco	774
Mozambique	20
Niger	124[b]
Nigeria	140
South Africa	173312
Swaziland	166
Tanzania	3
Zaire	110[a]
Zambia	471
Zimbabwe	3114
North America	
Canada	60480
Mexico	9771
USA	801600

Table 1.3 — contd.

Country	*Production (× 1000 tonnes)*
South America	
Argentina	400[a]
Brazil	7 800[a]
Chile	1 370
Colombia	9 706
Peru	130[a,b]
Venezuela	50[a]
Asia	
Afganistan	170[a,b]
Burma	43[a]
China	847 000
India	157 022
Indonesia	1 492
Iran	1 300[a]
Japan	16 383
Korea, North	70 000[a]
Korea, South	24 543
Mongolia	5 431[b]
Pakistan	2 135[b]
Philippines	1 294
Taiwan	1 858
Thailand	5 149
Vietnam	4 900[b]
Australasia	
Australia	175 498
New Zealand	2 409[a]
World Total	4 422 000

[a] Estimated figure.
[b] 1984 figure: 1985 figure not available.

Chapter 2

Mining Techniques and Sources of Environmental Problems

The types of pollution discussed here relate to relatively few processes (Fig. 2.1); however the precise nature of individual incidents depends upon the relative intensities of each component and how they are combined. In one consideration of environmental effects, the mineral mining industry was described as consisting of five stages (Ripley *et al.*, 1979):

Exploration — the geological, geochemical and geophysical surveying of an area to delimit ore bodies, plus an exploratory stripping or excavation
Development — the preparation of the mine for production, including the building of access roads and surface facilities
Extraction — ore removal activities
Beneficiation — concentration of the ore from low or medium grade deposits. This usually occurs near the mine site and results in the removal of the ore from gangue
Processing — carried out at any distance from the mine.

The first three stages occur in succession; however, once the mine is operational extraction, beneficiation and processing will all occur at the same time. Similar stages are associated with coal mining, although the terminology may be different.

The types of environmental effect observed are also related to the method of extraction. The three main methods are drift mining, deep mining and strip (or open-cast) mining; however the difference between deep and drift mining is now blurred as some 'deep' mines can be accessed by inclined drifts. The main distinction now is made between underground and surface mining, with the latter preferred on economic grounds except where there is a deep layer of overburden.

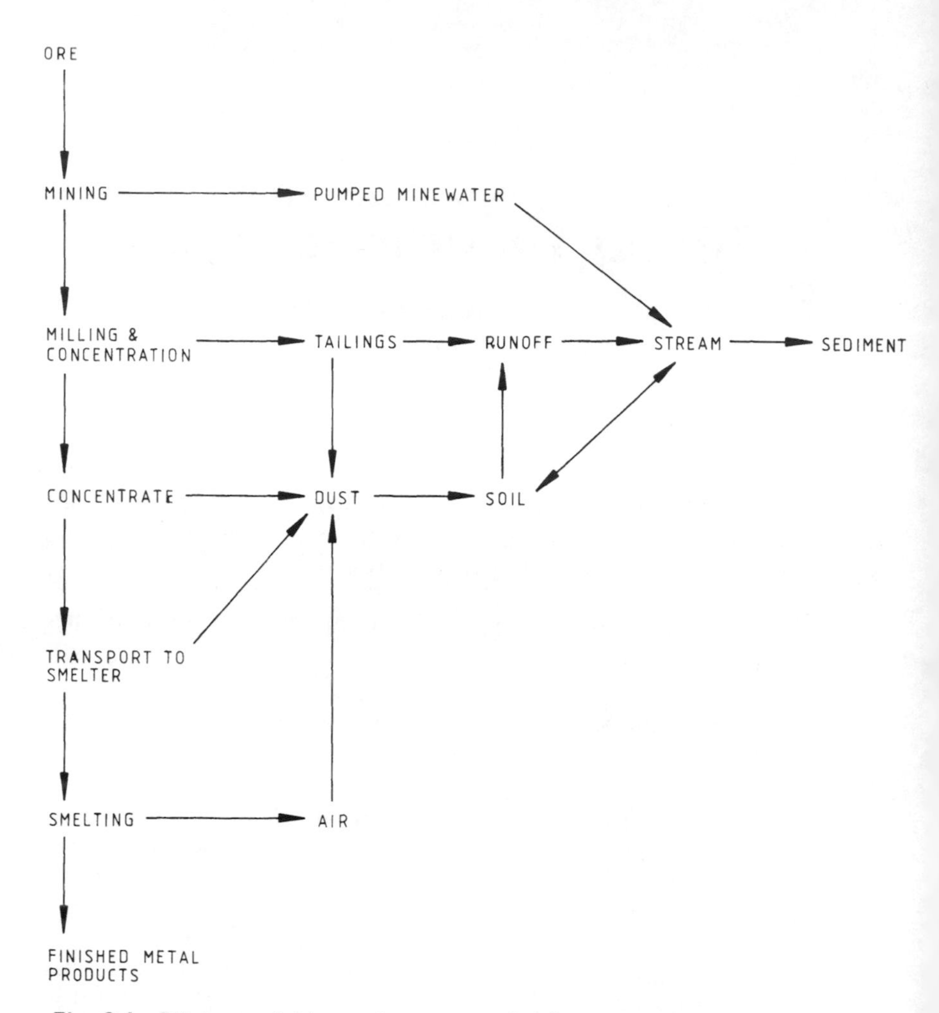

Fig. 2.1. Mining activities and transport of pollutants to the aquatic environment. This is a highly idealized diagram which outlines some of the major pathways of metals, other than direct discharges, to the aquatic environment.

DRIFT MINING involves exposing a horizontal or near horizontal seam on one of its sides into which the drift mine shaft (or 'adit') is driven (Fig. 2.2). The seam may slope up- or downwards from the point of entry. Seams sloping upwards will drain out through the entry shaft whereas those sloping downwards tend to fill with water.

Historically, this was a natural extension of open-pit and contour mining (see below) on hillsides. Exposed seams (of coal) or veins (of metals) were first exploited from the surface and the seams subsequently followed into the hillside.

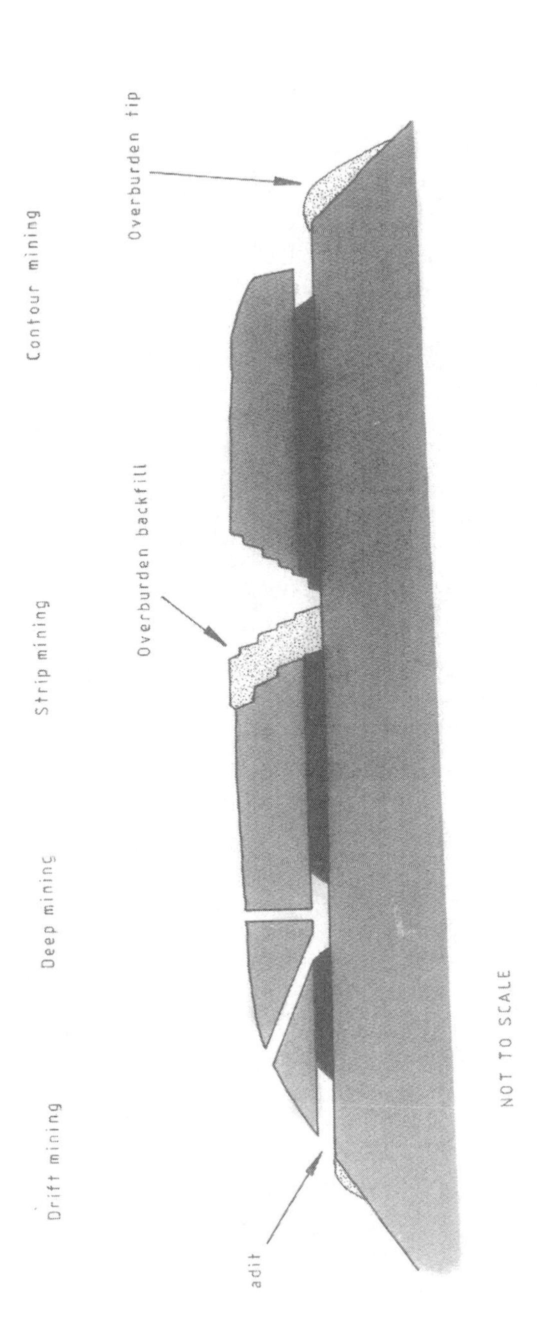

Fig. 2.2. Schematic representation of major mining techniques. In practice it is unlikely that a seam that can be strip mined will also be deep mined unless the thickness of the overburden increases significantly.

In DEEP MINING a vertical shaft is dug to intercept the mineral reserves (Fig. 2.2). Apart from excavation of the access shafts, no overburden material is removed. As the mineral reserves themselves are often below the water table, continuous pumping may be necessary to keep the mine operable. If the mine water is acidic (see below) then pumping will be more expensive as stainless steel pumps and plastic piping may be needed to resist the acid corrosion. Nowadays, however, plastic piping is often cheaper than metal anyway.

In some regions the bord and pillar technique for extraction of coal is common. Tunnels ('bords' or 'rooms') are criss-crossed throughout the seam leaving pillars between which support the roof of the seam (Watson, 1983). In Europe, however, mines almost exclusively use longwall techniques where a roadway in the plane of the seam is dug and automatic mining equipment placed along its length. This then works sideways into one of the walls of the roadway. As the mining proceeds the effect is to widen the original horizontal shaft but as the coal is removed only the roof adjacent to the longwall is supported, allowing the rest of the roof to collapse and fill the void left by the removed coal (Skelly & Loy, 1973).

STRIP (or OPEN-CAST) MINING is a widely-used technique where the mineral is close to the surface and occurs as a bedded deposit such as a coal seam. The overburden is removed in strips and placed into areas from which the mineral has already been removed (Fig. 2.2). Whilst mining is in progress this is used to reduce the visual impact of the operations and is returned when the operation is finished. The last strip mined may not have the overburden returned to it, in which case it may fill with water to form a strip-mine lake. Development of draglines to replace shovels has allowed easier and deeper extraction which, in turn, leads to larger and deeper lakes (Riley, 1960).

OPEN-PIT MINING is a surface mining technique similar to strip mining and is used where the mineral deposit is more irregular than that associated with strip mining.

CONTOUR MINING is a cross between strip and drift mining. The outer edge of the seam is revealed by stripping off the overburden along the contours of the hillside. Once this has been removed, the remaining seam can be exploited using a drift mining technique (Fig. 2.2). A number of variations of contour mining have been developed for coal mining; one of the more important is AUGER MINING, where a large mechanical auger bores into the seam, cutting the coal and bringing it out of the bore hole.

In Fig. 2.1 the interrelationships between major processes which cause pollution are outlined in a highly idealized form. Water is both a vital raw material for and a major waste from several of these processes and some of the direct discharges of water have been omitted from Fig. 2.1 for the sake of clarity. Much of the net influx is from natural sources, groundwater percolating through the rocks or draining in from surface runoff and rainfall; however under some circumstances it may be pumped in to reduce mine dust. Water pumped out often contains appreciable amounts of the mineral and associated trace metals, as well as spillage from underground repair and equipment maintenance.

Generally, metal ores are crushed underground and hoisted to the surface in skip loaders. Part of the water pumped from the mine to the surface is used in the milling procedures. Milling tends to produce fine waste, or tailings, which are allowed to settle out in tailing ponds before the water is discharged.

Processing the ores produces by far the largest amount of contaminated water. Chemical reagents may be added to separate the minerals from the finely ground rock by flotation. The variable mineral content of the ore may cause excessive use of reagents with poor quality ores, or low recovery of metals during pulses of very rich ores. From the flotation process the mixture is pumped into thickeners where the mineral concentrate is allowed to settle, is vacuum dried and then transported to smelters for final conversion to metal. The water, excess flotation reagents and colloidal and supercolloidal minerals are then discharged as a milling effluent.

The water and tailings waste from the mining and milling operations are discharged into settlement and treatment lagoons, termed ‘tailings ponds’. Dams for the ponds are usually constructed from tailings (generally 200 mesh in size), which are centrifuged into fine and coarse fractions during the ore concentration process. After the dams are constructed, the tailings and other liquid wastes are introduced upstream until the lagoon is effectively filled. The lagoons serve as settling basins for the residual tailings and biologically degrade spent organic reagents discharged from the milling process. Poor design, such as insufficient size or retention time, may cause problems with tailings ponds. Water may be lost into the ground, to cause aquifer pollution (Mink *et al.*, 1972) or through the retaining walls of the impoundment.

Concentrated ore may wash directly into stream systems from improperly placed storage piles or may be blown onto soil (up to 3–4 km downwind) and enter indirectly via runoff. Spoil heaps may contain considerable amounts of metals which were uneconomic to extract but which may be leached into the water for many years after mining has ceased. Pyritic minerals in the waste may acidify rainwater which would aid this leaching of toxic ions. Microorganisms may also play a role (Galbraith *et al.*, 1972).

Two final sources of pollution are transport emissions, material blown from uncovered lorries carrying concentrates, and smelter emissions, although there may also be visual and aural effects, ground vibration (due to blasting, mainly at surface mines) and subsidence of land over underground mines.

Different types of pollution are associated with coal mining. These commonly involve acidity (see Chapter 4) and suspended solids and, occasionally, heavy metals. Surface mining for coal is nearly always on a larger scale than deep mining. The largest strip mines extract up to 40 000 tonnes per day compared with 6000 tonnes per day for underground mine (Eddlemon & Tolbert, 1983). Excavation on this scale results in large piles of spoil and overburden which are often prone to erosion, releasing large amounts of sediment, acid and toxic leachates into watercourses (Riley, 1960) unless there are runoff ponds and drains to prevent this. Strip mines are also more likely to intercept surface streams and precipitation due to their larger surface area. In a survey of Canadian mines, Watkin (1983) estimated a mean flow of water of about 13 800 litres min^{-1} into open pits compared with about 1000 litres min^{-1} into active underground mines. Despite this, the majority of acid mine drainage problems do not originate from surface mines but from active or abandoned undergound mines (Skelly & Loy, 1973; Letterman & Mitsch, 1978). Nor is this problem common to all underground mines; of 7000 active mines in the USA only 2600 drain acid water (Lovell, 1983). This is partly a geological effect; for example in the USA lower concentrations of sulphur in coal and more alkaline materials in the overburden combine to reduce acid mine drainage problems (Olem, 1981; Grim & Hill, 1974).

Other workers have suggested that modern techniques may be exacerbating problems in some areas. Automation is resulting in less selective underground coal mining; more rock and debris is mined with the coal to be rejected after processing. This debris can account for up to half of the material extracted (Schlick & Wahler, 1976). As pyrite is especially common at the edges of seams, coal spoil may contain enough to produce acid mine drainage (Lovell, 1983). Complete removal of coal is virtually impossible, so some coal (and associated pyrite) will remain in the ground and this can continue to produce acid mine drainage (Watson, 1983).

Lower percentages of extractable material are often associated with metal ore, also frequently found as sulphides (Table 4.1). Mining these results in much larger quantities of spoil: up to 99·6% of the extracted material for copper may end up as processed waste (or tailings). Mining for soil shales in the USA is estimated to require removal of up to 190 000 tonnes per day to be economic. This would result in up to 220 000 m^3 day^{-1} of spent shale, often containing some pyrite in addition to potentially toxic hydrocarbon compounds (Eddlemon & Tolbert, 1983).

Chapter 3

Heavy Metals in the Aquatic Environment

3.1 METAL SPECIATION IN FRESHWATER SYSTEMS

One point that will be repeated over and over again in later chapters is that the speciation of a metal, rather than its total concentration, is the key to understanding its effects on the biota. The realization of this was due as much to a breakthrough in analytical techniques as it was to advances in theoretical knowledge of the subject; the technology for distinguishing different species of heavy metals at low concentrations has only become widely available relatively recently.

At its simplest, speciation distinguishes between 'filtrable' and 'particulate' fractions of a metal, two operational terms which separate that part of a water sample trapped by a filter and that which passes freely through (usually) a pore of 0·45 μm. Particulate material, by this definition, includes not only solid minerals and crystals of the metal but also metals adsorbed onto humic acids and other surfaces and incorporated into organisms. The filtrable fraction includes free ions and their complexes with various ligands (Fig. 3.1). The choice of pore size will determine ultimately what passes through; certainly some small particulates and colloids can pass through a 0·45 μm filter but this is offset by the ease and speed with which water samples can be filtered. Smaller pore sizes may achieve better separation but are correspondingly more difficult to use.

More detailed information on speciation requires access to a range of chemical techniques, typically including electrochemical methods (e.g. anodic stripping voltammetry (ASV) and ion-specific electrodes), electrodialysis techniques, phase separation techniques (see below) and some forms of chromatography, in addition to atomic absorption spectrophotometry (Stiff, 1971; Florence, 1977; Jensen & Jørgensen, 1984).

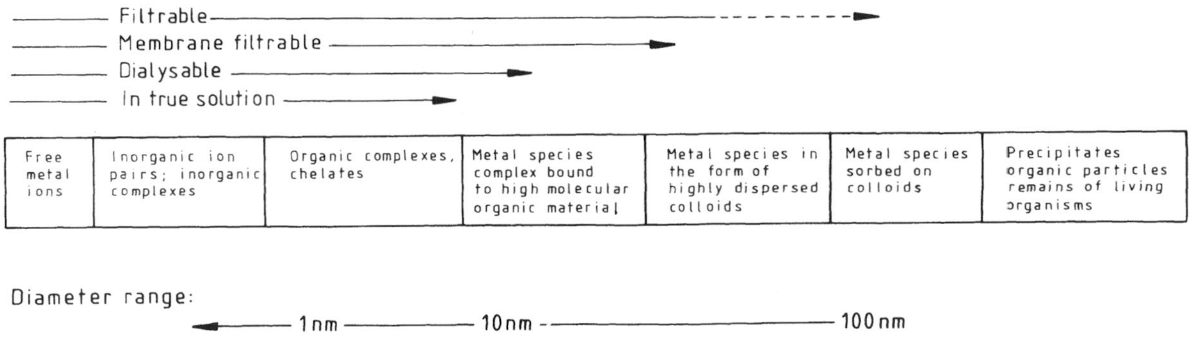

Fig. 3.1. Forms of metal species in natural waters. (Modified from Stumm & Bilinski (1973), and reproduced from *Advances in Water Pollution Research* with permission from Pergamon Press, Oxford.)

Although some electrochemical methods show great potential as methods for measuring the concentrations of ions directly (Astruc, 1979), they are complicated by poor sensitivity (ion-specific electrodes) or uncertain responses in low ionic strength solutions (anodic stripping voltammetry; Davison *et al.*, 1987).

If the chemical composition of the particulate phase is of interest, phase separation procedures (e.g. Gibbs, 1973), many of which were developed by workers interested in metal concentrations in sediments, are available. Typically, chemicals are used which selectively remove metals from different solid components to allow separation into those parts which are exchangeable, bound to carbonates, iron and/or manganese oxides or organic matter or more tightly bound in the mineral matrix (e.g. Gupta & Chen, 1975; Campbell *et al.*, 1979; Lion *et al.*, 1982); however incomplete knowledge of the complicated chemistry of the sediments coupled with the lack of selectivity of some reagents can make interpretation difficult at times (Tipping *et al.*, 1985; Tessier & Campbell, 1987). However Tessier and Campbell (1987) concluded that the selectivity of these procedures was generally acceptable within oxic surface sediments.

A complementary line of research has been the theoretical modelling of metal speciation, based on measurements of thermodynamic quantities such as stability constants (see Stumm & Morgan, 1981). Preliminary measurements made on simple solutions may later be combined into models based on mixtures of solutions resembling natural waters. A final step, the addition of particulate material, is often of great importance, especially in models of freshwater systems (Sibley & Morgan, 1975; Fontaine, 1984). Amongst the dissolved species in freshwaters, hydroxides, some carbonates and free ions are all important, compared with seawater where, not surprisingly, the chlorides dominate (Sibley & Morgan, 1975). It is important when using these models to accept their limitations. Sibley and Morgan (1975) appreciated that they did not incorporate biological effects into their model; however one other point to bear in mind is that the models ultimately rely upon experimental data, with all of the limitations discussed above. Failure to include all of the important factors, coupled with the assumptions inherent in every model, means that these methods probably provide only a rough estimate of the true situation (Wilson, 1976).

3.2 AQUEOUS ENVIRONMENTAL CHEMISTRY OF METALS

3.2.1 Nickel

There is less information available for nickel than for the other metals discussed in this book and most of what is available on its speciation is based on theoretical studies. However several workers have measured total and 'filtrable' concentrations in the field.

Typical concentrations are very low; Stokes (1981) quotes concentrations from < 0·002 to 0·010 mg $litre^{-1}$ whilst other workers report median values of 0·010 mg $litre^{-1}$ with ranges from 0 (*sic*) to 0·071 mg $litre^{-1}$ (Durum & Haffty, 1963) and 0·0002 to 0·020 mg $litre^{-1}$ (Bowen, 1966). These figures are very similar to the mean concentration of 0·015 mg $litre^{-1}$ (total nickel; range = 0·003 to 0·070 mg $litre^{-1}$) reported from a survey of 105 stream sites in Northern England (Wehr & Whitton, 1983*b*). At the other extreme, highly contaminated streams receiving leachate from the nickel smelter at Sudbury, Ontario contained up to 6·4 mg $litre^{-1}$ (Stokes *et al.*, 1973).

In most natural waters (pH 5–9) the predominant species are either the free ion, Ni^{2+} (Morel *et al.*, 1973; Vuceta & Morgan, 1978; Richter & Theis, 1980), the carbonate (Sibley & Morgan, 1975) or a mixture of both (Mouvet & Bourg, 1983), along with other species such as hydroxide, sulphate, chloride and ammonia (Morel *et al.*, 1973). Under the reducing conditions that are sometimes found the low solubility of nickel sulphide (log Ks = −22·9) may limit dissolved nickel concentrations (Morel *et al.*, 1973; Richter & Theis, 1980).

A large proportion (5–98%) of the total nickel concentration is often in the non-filtrable fraction, probably representing nickel adsorbed onto particulate material (Jenne, 1968). Speciation studies in the River Meuse (Mouvet & Bourg, 1983) showed that this, rather than precipitation, probably controlled the concentration of dissolved nickel in this, usually aerobic, river.

3.2.2 Copper

In contrast to nickel, copper can form stable compounds in more than one oxidation state (Cotton & Wilkinson, 1972). Copper(II) is the normal oxidation state for soluble aqueous complexes although insoluble complexes of copper(I) are also quite stable. Copper(III) complexes are relatively few in number and are unstable in aqueous media (Leckie & Davies, 1979).

Copper(I) is classified as a soft acid and forms stable complexes with typical soft bases (iodides, sulphides, thiosulphates, etc.); copper(II) lies between the hard and soft acid classifications and can form complexes with either, with a preference for the former (carbonates, sulphates, hydroxides, chlorides, etc.) (Pearson, 1968).

The relative stabilities of copper(I) and copper (II) species in solution varies considerably depending upon the nature of the ligands present.

Soluble copper(I) compounds are stable only at very low equilibrium concentrations. Stable insoluble copper(I) compounds include copper(I) sulphide, copper(I) chloride and copper cyanide; the halide and cyanide complexes are actually more stable in water than the copper(II) complexes. Nonetheless it is the chemistry of copper(II) which is of the greatest interest to freshwater biologists as a great many of its salts are water-soluble.

A vast number of measurements of filtrable copper have been reported and any summary here is necessarily selective. Some typical estimates for (relatively) unpolluted water include a mean of 0·005 mg litre^{-1} for 'average river water' (Riley & Chester, 1971), a median of 0·010 mg litre^{-1} with a range of 0·0006 to 0·40 mg litre^{-1} (Bowen, 1966), a median of 0·0053 mg litre^{-1} with a range of 0·0008 to 0·105 mg litre^{-1} for large rivers in North America (Durum & Haffty, 1963), an 'approximate estimate in streams' of 0·007 mg litre^{-1} (Turekian, 1971) and concentrations generally less than 0·005 mg litre^{-1} in Lake Kinneret, Israel (the biblical 'Sea of Galilee'; Frevert, 1985). Relatively few areas are now far enough away from industrialization to represent genuine 'background' concentrations; concentrations in the Amazon and its tributaries ranged from 0·00038 mg litre^{-1} to 0·00235 mg litre^{-1} (Boyle, 1979).

There is a general consensus in the literature that under normal conditions most copper(II) is adsorbed and the concentration of the free ion is very low (Boyle, 1979; Wagemann & Barica, 1979). The concentration in the filtrable fraction may be much higher (e.g. 3–80%; Wilson, 1976) but much of this may be in the form of soluble complexes rather than as free ions. Processes which may control the concentrations of free copper include hydrolysis (Pagenkopf *et al.*, 1974; Gachter *et al.*, 1978), precipitation (Sylva, 1976; Leckie & Davies, 1979) and, especially, complexation (Stiff, 1971; Millero, 1975; Sylva, 1976; Mantoura *et al.*, 1978; Mouvet & Bourg, 1983) particularly with organic compounds (Bowen, 1985). Some of the forms found or predicted are listed in Table 3.1.

The tendency for copper to adsorb, is particularly strong when the surfaces of particulates are modified by organic or manganese and iron oxide coatings (Lion *et al.*, 1982; Tipping *et al.*, 1983). Manganese and iron oxides are particularly suitable substrates as copper (and other metals) can accumulate both by direct adsorption (Johnson, 1986) and by co-precipitation as the crystals grow (Laxen, 1984).

3.2.3 Zinc

The chemistry of zinc is a little easier than that of copper as only one

Table 3.1
Copper forms found or predicted in freshwater. Most of these forms will also apply to other metals, although copper is particularly strongly associated with organic complexes

Form	*Comments*	*Reference*
$CuCO_3$, CuCN Complexes with amino acids, polypeptides and humic substances	In polluted surface waters, some with sewage	Stiff (1971)
Free ion Complexed by carbonate and hydroxide ions	Alkalinity is factor controlling free ion concentration	Pagenkopf *et al.* (1974)
$Cu(CO_3)_2^{2-}$, $CuCO_3^{\circ}$, 1% as free ion	Derived from equilibrium models. Sensitive to the pH and chemical composition of the water	Millero (1975)
Precipitation as malachite	Kinetics of precipitation may be limiting	Sylva (1976)
$CuCO_3$	Present at higher pH values	
Mainly associated with organic matter, probably organic colloids	Sequential analysis procedure adopted	Florence (1977)
Complexes of carbonates and hydroxides plus organic complexation		Gachter *et al.* (1978)
Cu–humic acid complex	Lake Celyn, North Wales	Mantoura *et al.* (1978)
Organic complexes and adsorption	River Meuse, Belgium	Mouvet & Bourg (1983)

oxidation state, zinc(II), is stable. In terms of its chemical interactions with ligands it is intermediate between the hard and soft acids. More practically, because of its greater solubility and generally higher concentrations it has been somewhat easier to study than some of the other metals discussed here. On the debit side, the widespread use of zinc makes contamination a very real problem, particularly when studying low concentrations (Martin *et al.*, 1980).

Notwithstanding this, some very low concentrations have been reported for rivers draining generally undeveloped regions. Shiller and Boyle (1985) measured filtrable concentrations in these areas between 0·00002 and 0·0018 mg $litre^{-1}$. In rivers in North America where some addition of zinc from anthropogenic sources is expected, they measured concentrations between 0·00007 and 0·0156 mg litre^{-1}. 'Clean' streams in North Wales contained 0·011 mg litre^{-1} in one study (Abdullah & Royle, 1972) and up to 0·075 mg litre^{-1} (mean = 0·024 mg litre^{-1}) in another (Elderfield *et al.*, 1971); however when sites below specific discharges are examined then much higher concentrations may be found. Typically these are often above 0·100 mg litre^{-1}, sometimes exceed 1·00 mg litre^{-1} and can be much higher. For example, Say and Whitton (1982) measured 42·5 mg litre^{-1} in the filtrable fraction of water draining a lead mine in the French Pyrenees.

In the pH range of most natural waters (*ca* 6–8) the free ion is the predominant form (Hem, 1972; Vuceta & Morgan, 1978). Above this carbonates and hydroxides become the dominant species although these are also quite soluble and thus cannot control zinc solubility except at very high concentrations of dissolved carbon dioxide (Hem, 1972). Willemite (zinc silicate) is less soluble than these but low concentrations of dissolved silica generally preclude this from acting as a significant control on solubility. Under these conditions it is adsorption rather than precipitation which is the key factor controlling zinc concentrations and this too is strongly affected by pH, with virtually no adsorption at pH 6 (Florence, 1980) and increasing adsorption at higher pH values (Shiller & Boyle, 1985; Johnson, 1986). Despite this, zinc has a lower tendency to adsorb than copper (Vuceta & Morgan, 1978) and a greater proportion is likely to exist in the aquo form.

3.2.4 Lead

Lead differs from the other three metals considered here in that it is not a transition element, but rather a member of Group IV along with carbon, silicon, germanium and tin. Like copper it is intermediate between hard and soft acids in its interactions with ligands and has more than one stable oxidation state. It is also the densest of the four metals (atomic mass = 207·2). As it is the metallic properties which are of prime interest here the differences between Group IV elements and transition elements need not be elaborated upon except to say that Group IV elements are capable of forming organoderivatives which are economically very important. These constitute an important environmental source of lead.

Background concentrations of lead reported in the literature include 0·000006 to 0·000050 mg litre^{-1} measured in remote streams in the USA (Settle & Patterson, 1980; measured by mass spectrometry). However these concentrations are two orders of magnitude lower than the minimum concentrations reported for unmineralized streams in North Wales (0·0005 mg litre^{-1}, Elderfield *et al.*, 1971; 0·0007 mg litre^{-1}, Abdullah &

Royle, 1972) and three orders of magnitude lower than background concentrations in rivers in the New Lead Belt in Missouri, USA prior to mining (0·004–0·006 mg litre^{-1}, Wixson, 1972). Concentrations in rivers affected by mining are generally at least a couple of orders of magnitude higher than these (e.g. 0·065 mg litre^{-1} in River Derwent, Co. Durham; Harding & Whitton, 1978). Concentrations greater than 1 mg litre^{-1} are rare, although, based on equilibrium calculations, they are theoretically possible (Laxen & Harrison, 1983). Say and Whitton (1982) measured 1·11 mg litre^{-1} at Estaing Mine, French Pyrenees.

It is the lower oxidation state, lead(II), which is the most stable under normal, oxidizing, conditions. Below pH 7·1 it is present mainly as the free ion and above this as the carbonate and hydroxide (Vuceta & Morgan, 1978). At pH 7·5 the carbonate is assumed to act as the major controlling solid (Jørgensen & Jensen, 1984); the hydroxide is insoluble only above pH 10·0 (Moore & Ramamoorthy, 1984). At acid pH values the sulphate may control solubility (Laxen & Harrison, 1983). As for the other metals adsorption plays an important role in controlling concentrations in solution.

There is one feature of the aquatic chemistry of lead which is not a feature of the other three metals. Under anaerobic conditions in the laboratory methylation of lead has been demonstrated, both chemically (Jarvie *et al.*, 1975) and biologically (Wong *et al.*, 1975; Schmidt & Huber, 1976) mediated. More recently the biologically-mediated process (biomethylation) has been shown to involve the transfer of methyl groups from vitamin B_{12} to the heavy metals (Wood & Wang, 1985). Moore and Ramamoorthy (1984) emphasized that so far this has only been demonstrated in the laboratory and care should be exercised in extrapolating these data to natural systems. However, because organometallic species (such as methyl mercury) are often more toxic than inorganic species of the same metal, a more thorough understanding of the environmental chemistry of this process would be highly desirable.

3.3 TRANSPORT, DEPOSITION AND CYCLING OF METALS

It is clear from the discussion above that much of the dissolved metal which enters rivers is going to be sorbed onto particulates (Nienke & Lee, 1982) or, under certain conditions (e.g. high alkalinity or pH), precipitate out (Jennett & Foil, 1979). Along with this much metal-rich dust and particulates from smelters, tailings ponds and effluents may undergo little or no change after entering a river. The average residence times of these particles in rivers (of the order of days or weeks) is too short for the establishment of stable, dynamic equilibria between the water and suspended material (Bowen, 1975). In a study of transition metal transport from the rivers Amazon and Yukon to oceans Gibbs (1977) concluded that less than 3% was associated with dissolved species. In another study the ratio of solid to aqueous phases was between 2:1 and 1:1 for zinc and between 4:1 and 2:1 for lead (Forstner & Prosi, 1978).

Once the metal is associated with the solid phase then it can be transported for great distances so long as the critical velocity of the river is high enough to keep it in suspension. In the South Esk River in Tasmania, the concentrations of copper in the water, suspended solids and sediment exceeded background concentrations for 130 km downstream of the source of the contamination (Norris *et al.*, 1981). The total amount of metals which may be transported in this way is considerable, the Susquehanna River, in the south-eastern USA, transports 88·4 $t\,y^{-1}$ chromium, 45·7 $t\,y^{-1}$ silver and 3048·2 $t\,y^{-1}$ nickel (Turekian & Scott, 1967) and the Grand Calmut River discharges 84 $t\,y^{-1}$ zinc, 1·2 $t\,y^{-1}$ cadmium and 24 $t\,y^{-1}$ lead into Lake Michigan (Romano, 1976). The Mississippi River was estimated to transport 2100 t of lead to the Gulf of Mexico between 1982 and 1983 compared with 3600 t between 1974 and 1975, a decrease at least in part attributable to the decline in use of tetraethyl lead as an additive in petroleum (Trefry *et al.*, 1985).

As the flow regime of a river changes so the concentration of suspended metal will also fluctuate. In the River Ystwyth, mid-Wales, metal concentrations in suspended material were lowest during periods of high flows as a result of dilution of the river water by rainfall and runoff (Bradley & Lewin, 1982). As discharge began to decline to normal, however, sorption and co-precipitation by manganese and iron complexes appeared to act as metal 'sinks'. When the 'total' metal concentration is considered, different effects are apparent. In the River Hayle in Cornwall, for example, copper concentrations were lower during periods of low flow in the summer whilst concentrations of zinc were higher (Brown, 1977*a*). However 80% of the zinc was in the 0·45 μm filtrable fraction and so fluctuations in the suspended load do not, perhaps, have the same effect as they may have on copper concentrations. This hypothesis is supported by results of Hart *et al.* (1982) working on Magela Creek in Northern Territory, Australia. Here there was little variation in the filtrable concentration with flow whilst the total concentration, which was closely correlated with suspended solid concentrations, increased at high flows. The peak concentrations occurred before the peak in discharge suggesting that the first runoff water contained the bulk of the erodable material. Relationships between flow rates and metal concentrations are also affected by concomitant changes in runoff patterns and pH (Bird, 1987).

Changes in the flow of a river may lead to the settling or resuspension of particulate material. Thus the sediments may contain higher concentrations of heavy metals at points where the river flow is particularly low (Mudroch & Capobianco, 1980). The sediments and benthos may also act as sinks by sorbing metals from the water. This was elegantly demonstrated by Kuwabara *et al.* (1984) in a study of a stream in the Sierra Nevada, California. They added copper (as copper sulphate) and chloride (as sodium chloride) to the stream as a square wave pulse over 9 h and measured concentrations 400 m and 700 m downstream. For chloride, a biologically 'conservative' element, concentrations at the two sites were very similar; however for copper there was marked attenuation at the downstream site which they were able to attribute to sorption both by sediment and periphyton.

Major changes in flow occur when rivers enter lakes. The decrease in velocity may result in deposition of the suspended material. Contamination of bottom sediments by copper, zinc and lead from domestic sources decreased with distance away from an input into Lough Neagh in Northern Ireland (Rippey, 1982) and similar patterns were seen in studies of Ullswater in the English Lake District below the inflow of a stream containing mine wastes (Denny, 1981) and the Derwent Reservoir in Northern England (Harding & Whitton, 1978). Concentrations of zinc, cadmium and lead in this case were greater at the end into which the mine effluent-containing river entered (Fig. 3.2). Here the sediments acted as a major sink for heavy metals and the water which is abstracted at the other end of the lake contains concentrations which are considered acceptable for subsequent treatment for domestic use (Wallwork & Hunter, 1981). A similar situation occurred in Lake Kinneret in Israel where the concentrations of zinc and lead, both in dried sediment and in interstitial water, followed the path of the River Jordan through the lake (Frevert & Sollman, 1987).

For both rivers and lakes the nature of the sediments is an important factor controlling their incorporation of metals. Adsorption and precipitation are functions of surface area and the capacities of sediments for these processes will increase as the organic content increases and the particle size decreases (McDuffie *et al.*, 1977; Ramamoorthy & Rust, 1978; Sakai *et al.*, 1986). Norris *et al.*, (1981), in their study of the South Esk River, found

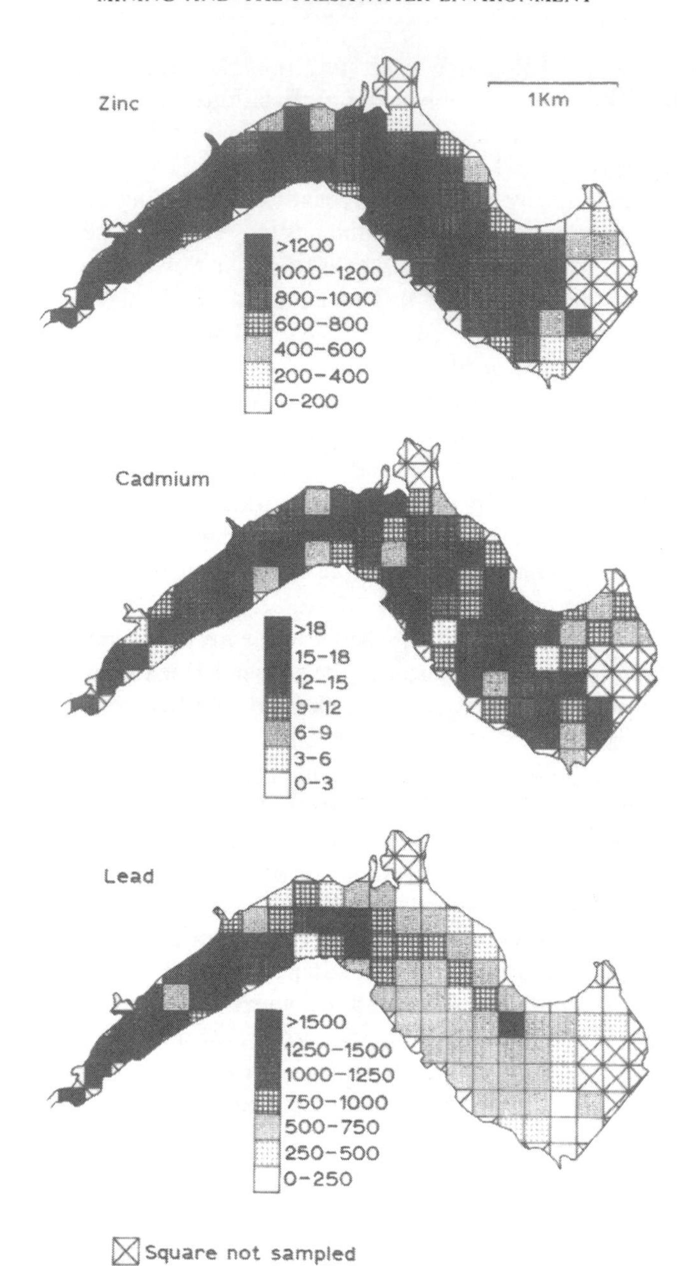

Fig. 3.2. Variations in zinc, cadmium and lead concentrations (in $\mu g\, g^{-1}$) in Derwent Reservoir sediments. Note higher lead content following the original course of the River Derwent. (From Harding & Whitton (1978), *Water Research* **12**, 307–16 and reproduced with permission from Pergamon Press, Oxford.)

that where there was little retention of fine clay and humic material then the concentrations of heavy metals in the sediments was low. Likewise, in a study on Lake Ontario, Nriagu *et al.* (1981) found organic material to be an important vehicle for metal transport to lake sediments; however Moriarty *et al.* (1982), who studied the River Ecclesbourne in Derbyshire, found that most of the heavy metals from a mining site were transported and deposited as the original mineral particles. These three very different results show how the chemical phase of the effluent influences its subsequent behaviour.

If lake sediments act as the eventual 'sink' for metals then cores of sediments will contain a record of heavy metal inputs to the lake as they do for pollen grains. Furthermore studies on these metals can benefit from the advances made by palaeobotanists on operational difficulties such as the effects of vertical mixing at the sediment–water interface. Analyses of this kind can be used to establish base line concentrations, identify temporal changes in atmospheric inputs and assess the effects of specific inputs within the catchment. As the relation of sediment to water concentrations is often difficult most workers have preferred to use sediment cores to pick out changes in deposition pattern and relate these to historical records. Under ideal conditions the record may extend back well before the industrial revolution.

One such case is Blelham Tarn, a small (0·1 km^2), productive lake in the English Lake District with no specific inputs of heavy metals. Here, concentrations of copper, zinc, cadmium and lead all showed enhancement after the onset of the Industrial Revolution which, in the case of copper, zinc and cadmium, accelerated during the 20th century (Fig. 3.3). Lead, which was mined in the vicinity, showed a more dramatic rise throughout the 19th century whilst the flux of nickel showed no such anthropogenic increase and instead mirrored changes in major cations, suggesting that its method of transport to the sediments was similar (Ochsenbein *et al.*, 1983). The second major cause of increases in lead concentrations is the use of tetraethyl lead derivatives as 'anti-knocking' agents in petrol and in sediments of Lake Michigan in North America these were attributed to marked increases in lead concentration since about 1920, although before this most of the anthropogenically-derived lead entering the lake came from the combustion of coal (Edgington & Robbins, 1976).

In a catchment where copper was mined (Coniston, English Lake District) analysis of sediment cores from throughout the lake allowed some estimation both of the magnitude of the pollution from the mines and the subsequent distribution of the copper throughout the lake

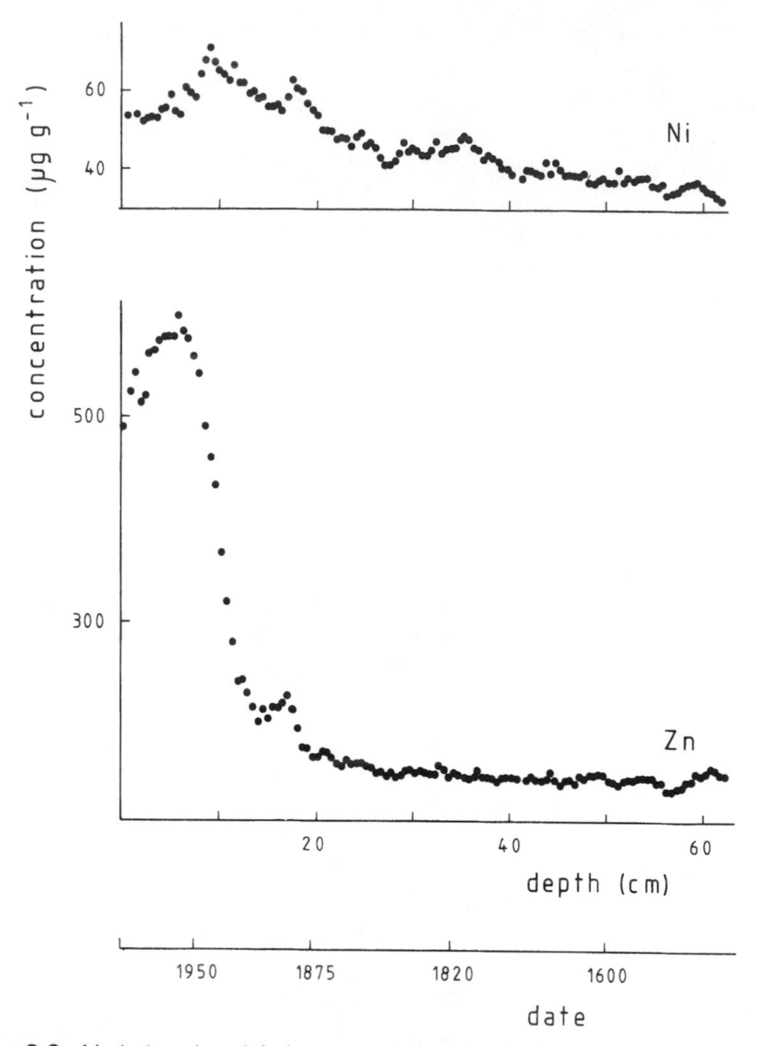

Fig. 3.3. Variation in nickel, copper, zinc and lead concentrations in a sediment core from Blelham Tarn, English Lake District. Two outliers for lead (380, 1290 $\mu g\, g^{-1}$) have been omitted. Approximate dates of deposition of different levels in core are derived from diatom stratigraphy. (Modified from Ochsenbein *et al.* (1983) and reprinted from *Archiv für Hydrobiologie* **98**, 463–88 with permission from E. Schweizerbart'sche Verlagsbuchhandlung.)

(Davison *et al.*, 1985). Although copper entered the lake through two inflows at the northern end of the lake all of the sediment cores showed pronounced copper peaks at the same point in the profile, albeit of

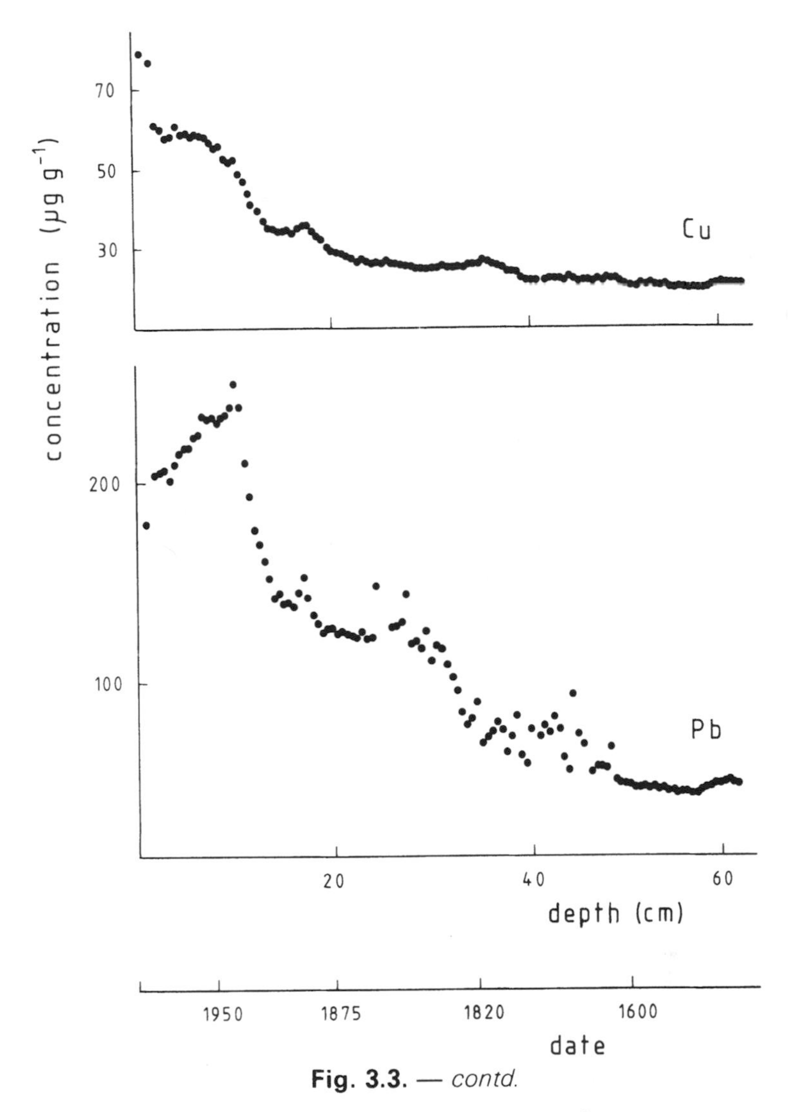

Fig. 3.3. — *contd.*

different magnitudes, indicating the importance of transport processes within the lake. As the copper to sulphur ratio was 2:1 at all sites they concluded that the copper was present in the sediments as particles of chalcopyrite ($CuFeS_2$), the main ore mineral, and did not undergo subsequent transformations (Davison *et al.*, 1985).

Superimposed on the physical and chemical processes, however, is the influence of the biota. In lakes (Fig. 3.4) the biological cycle is confined mostly to the upper water (the 'euphotic zone'). Phytoplankton and other

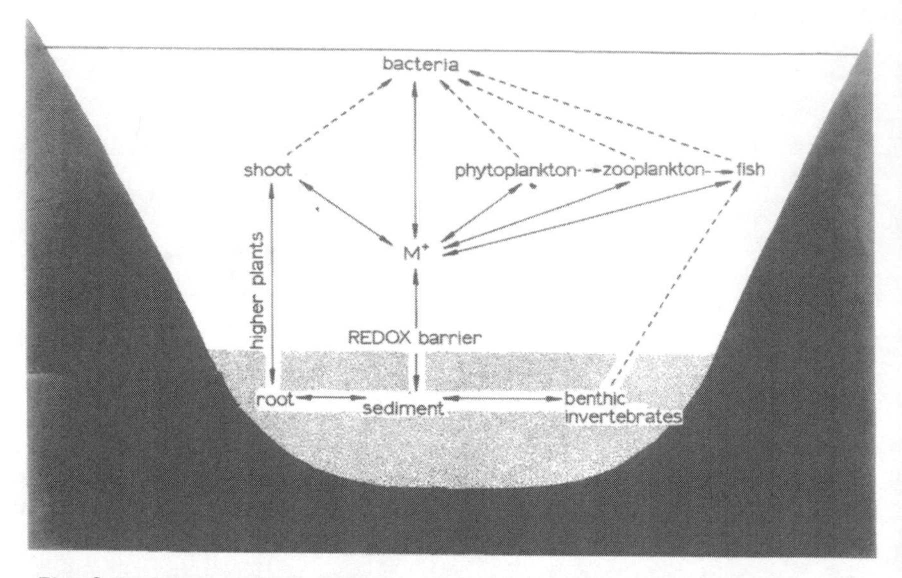

Fig. 3.4. Schematic representation of the various internal pathways of metal cycling in lakes. Dashed lines represent movement along food chains whilst solid lines indicate chemical pathways. In practice the role of the biota will decline in deeper parts of lakes whilst biogeochemical factors increase. See text for more details.

particulates take up trace metals (Baccini *et al.*, 1978). When they die the metals, both in dissolved and particulate forms, are released by bacterial decomposition (Jackson, 1978); in stratified lakes this may occur in either the epi- or the hypolimnion as the dead algae sink through the water column. Some of the particulate matter may be lost by sedimentation (Hamilton-Taylor *et al.*, 1984), although it may stay in suspension long enough to scavenge more metal (Sigg, 1985) whilst the dissolved metals may cycle again. In a study on Ullswater it was suggested that certain phytoplankters which overwinter on bottom mud (e.g. *Melosira*) may take up lead from this and carry it with them when they rise to the upper water column in the spring (Welsh & Denny, 1979). The lead may subsequently return to the sediments when the plankton bloom 'crashes'. Similarly the zooplankter *Mysis relicta* was shown to play an important role in transporting cadmium and zinc from the sediments to the water column of a lake in Ontario by its daily migrations and then make these available to the rest of the biota (Van Duyn-Henderson & Lasenby, 1986).

If the lake is shallow or is not thermally stratified, then some of the metal may be transported close enough to the sediments for direct adsorption to occur. A similar cycle will occur in fluvial systems although the relative importance of certain reservoirs and pathways will differ. In streams and shallow parts of rivers and lakes a benthic reservoir, composed of either algae or macrophytes, may also be important.

To regard the sediments solely in terms of metal sinks is clearly over simplistic. Metals 'lost' to the sediments may subsequently be released by biogeochemical cycling (Hamilton-Taylor *et al.*, 1984), by turbulence (Everard & Denny, 1985*a*) or by uptake by the biota (Welsh & Denny, 1976, 1980; Reynoldson, 1987). The former is perhaps most likely in deeper waters where there are pronounced redox cycles whilst the latter two are associated with shallower parts of lakes and, presumably, rivers as well. Pathways from the sediments back to water include uptake by macrophyte roots and along food chains from sediment-dwelling invertebrates, which often ingest large amounts of mineral material (see Chapter 5).

Under anaerobic and reducing conditions insoluble metal sulphides may form; in Lake Kinneret hydrogen sulphide builds up in the hypolimnion during summer stratification and this enhances the loss of metals from the water to the sediment as sulphides (Frevert, 1987). However, there were marked differences in the ease with which they were subsequently oxidized back to soluble forms with zinc being returned to solution almost immediately, copper requiring longer periods of oxic conditions, whilst only about 1% of lead was returned after seven days' continuous oxidation. On the other hand, during periods of deep mixing when the sediment surface is oxic then heavy metals may be lost to the sediments either adsorbed or co-precipitated onto the less-soluble forms to be released again when the sediment interface is anaerobic (Laxen, 1984; Gunkel & Sztraka, 1986).

Which of these is the dominant process in anoxic waters? Salomons *et al.* (1987) believe that it is the precipitation of sulphides, pointing to studies reporting the lack of an expected relationship between pore water concentrations and sediment concentrations expected if adsorption was the dominant process. In oxic waters and sediments, however, all sulphide is very quickly oxidized back to sulphate.

In other words, whilst sediments are the ultimate sink for heavy metals, the surface layers at least are still able to play a role in the cycling of metals within the lake.

3.4 CONCLUSION

Perhaps the firmest conclusion which may be made from this chapter is just how difficult it is to make generalizations about heavy metals in the aquatic environment. That is not to say that all of those made are wrong or that suitably qualified generalizations are not sometimes appropriate. What is dangerous is to apply these to new situations without very careful consideration of the idiosyncrasies of each. As, later in the book, interactions between metals and the biota are considered, one should be aware of the physical and chemical effects which may be influencing the results as reported.

Chapter 4

Acid Mine Drainage in the Aquatic Environment

4.1 MINING AND ACID PRODUCTION

Although any mineral deposit which contains sulphide is a potential source of acid mine drainage certain types of mining are more prone than others. There are records of acid drainage where coal, pyritic sulphur, copper, zinc, silver, gold, lead and uranium amongst others have been mined (Barton, 1978; Du Plessis, 1983). Coals and shales of marine origin tend to contain higher concentrations of sulphide than strata from freshwater palaeoenvironments (Eddlemon & Tolbert, 1983; Lovell, 1983). Some of the more important sulphide minerals are listed in Table 4.1.

Chemically, pyrite, the most important mineral, is iron disulphide (FeS_2). Marcasite, an orthorhombic polymorph of FeS_2, is more reactive than pyrites. Variations in pyrite morphology such as crystallinity, particle size and reactivity all affect its breakdown; in particular crystalline forms are less subject to weathering and oxidation than amorphic forms (Riley, 1960; Barnes & Romberger, 1968; Grim & Hill, 1974; Lovell, 1983).

The reactions involved in the breakdown of pyrite in the presence of water and oxygen to yield sulphuric acid are well known (Singer & Stumm, 1970):

$$FeS_2 + 7/2\ O_2 + H_2O \rightarrow Fe^{2+} + 2SO_4^{2+} + 2H^+ \quad (1)$$

$$Fe^{2+} + 1/4\ O_2 + H^+ \rightarrow FE^{3+} + 1/2\ H_2O \quad (2)$$

$$Fe^{3+} + 3H_2O \rightarrow Fe(OH)_2 + 3H^+ \quad (3)$$

$$FeS_2 + 14Fe^{3+} + H_2O \rightarrow 15Fe^{2+} + 2SO_4^{2+} + 16H^+ \quad (4)$$

It is clear from these reactions that the pyrites can remain in their reduced

Table 4.1
Some important mined sulphides

Mineral	*Composition*
Arsenopyrite	$FeS_2 \cdot FeAs$
Bornite	$CuFeS_4$
Bravoite	$(Ni, Fe, Co)S_2$
Chalcocite	Cu_2S
Chalcopyrite	$CuFeS_2$
Cinnabar	HgS
Cobaltite	$CoAsS$
Covellite	CuS
Enargite	$3Cu_2S \cdot As_2 \cdot S_5$
Galena	PbS
Gersdorffite	$NiAsS$
Millerite	NiS
Molybdenite	MoS_2
Orpiment	As_2S_3
Pentlandite (Mackinawite)	$(Fe, Ni)_9S_8$
Pyrite	FeS_2
Pyrrhotite	$Fe_{11}S_{12}$
Realgar	AsS
Sphalerite	ZnS
Stibnite	Sb_2S_3
Tetrahedrite	$3Cu_2S \cdot Sb_2S_3$
Violarite	$(Ni, Fe)_3S_4$

state in undisturbed strata so long as they are anaerobic. Whilst there are a few cases of naturally occurring acid streams, most occur as a result of mining activities.

Which is the rate limiting step has been the subject of some controversy. Whilst some workers considered it to be the reaction of FeS_2 and O_2 (eqn (1)) Singer and Stumm (1970) were able to show that it was the oxidation of the ferrous iron (iron(II) — eqn (2)) and that it was therefore irrelevant whether the FeS_2 was pyrite or marcasite. Note that there is a propagation cycle between eqns (2) and (4) where Fe^{3+}, one of the products of eqn (2), acts as an oxidant of the pyrite in eqn (4) and Fe^{2+} produced by eqn (4) can be used as a reductant in eqn (2). As the process is limited by the oxidation of pyrite the surface area available for oxidation determines the rate of the reaction (Gottschlich *et al.*, 1986). The chemistry of the reactions and the debate on the rate limiting steps are described in more detail in Barnes and Romberger (1968) and Barton (1978).

Bacteria which are able to utilize pyrites as an energy source act as catalysts and can increase the rate of oxidation by up to one million times (Singer & Stumm, 1970). In particular the acidophilic chemoautotrophs *Thiobacillus ferrooxidans* (iron-oxidizing) and *T. thiooxidans* (sulphur-oxidizing) have been found in virtually all cases of acid mine drainage. In spoil they may be found on the surface of the coal (Apel *et al.*, 1976) which they may attack with enzymes; however in streams, where they are more likely to be washed away, they are often embedded in white or cream coloured filamentous 'streamers' composed of a network of fibrillar polymers secreted by bacteria (Dugan *et al.*, 1970; Wakao *et al.*, 1985). These organisms keep the ratio of Fe^{3+} to Fe^{2+} in the Fe^{3+} oxidizing step (eqn (4)) high (Dugan, 1975) and their activity may be enhanced by the addition of surface wetting agents such as detergents (Sand, 1985). At least part of the energy requirements of these bacteria may be satisfied by energy generated by the natural hydrogen ion gradient across their cell membranes (Apel *et al.*, 1980).

This understanding of the role of bacteria in enhancing the degradation of pyrites has been coupled into industrial processes to treat coal prior to its combustion. Not only does this increase its energy and decrease its ash content but it also reduces the threat of acid precipitation which accompanies the burning of sulphur-rich coal (Dugan, 1986). Dugan (1984) suggested a system which, in outward design at least, resembled an activated sludge plant for sewage treatment; however in this case the aeration and mixing tank is seeded with mixed cultures of chemotrophic and heterotrophic bacteria (Dugan, 1984). On the other hand, addition of compounds such as sodium lauryl sulphate and benzoic acid to coal spoil had the opposite effect and inhibited the bacterial oxidation of pyrites (Dugan, 1987*a*, *b*).

One new analytical tool which has been applied to acid mine drainage waters is the analysis of isotopic composition of the oxygen in sulphate within acid mine drainage to distinguish between sulphate produced from oxygen derived from air, which has up to 18 parts per thousand of ^{18}O, and water, with a much lower proportion of ^{18}O (Taylor *et al.*, 1984*a*). As a result it is now possible to distinguish between sulphate produced in anaerobic, sterile environments where the sole source of oxygen is water and aerobic environments with chemotrophic bacteria where molecular oxygen is also used. This may prove useful in assessing the effectiveness of abatement programmes (Taylor *et al.*, 1984*b*; van Everdingen & Krouse, 1985). More detailed aspects of the biology of *Thiobacillus* have been reviewed by Lundgren and Silver (1980) and Harrison (1984).

4.2 FACTORS AFFECTING THE ENVIRONMENT

Acid mine drainage is very often a multi-factor pollutant and the importance of each factor will vary both within and between affected systems. The main factors — the acidity itself, ferric (iron(III)) precipitates, heavy metals and turbidity —will be considered separately.

4.2.1 Acidity

Although pH values are easy to collect and compare and most authors give details, pH is neither a true indication of acidity (Barton, 1978), nor is it a good determining factor for the extent of acid mine drainage (Moon & Lucostic, 1979). The streams with the lowest pH values are not necessarily those with the most depauperated biota (Carlson-Gunnoe *et al.*, 1983).

pH is an intensity factor, measuring the concentration (or, more strictly, activity) of hydrogen ions whereas what is most important in situations with acid mine drainage is not the concentration alone but the availability of hydrogen ions to neutralize bases; in other words, their excess over other ions. This quantity is usually referred to, perhaps rather confusingly for non-ecologists, as 'total acidity' and is reported in the same units as alkalinity (usually mg $litre^{-1}$ $CaCO_3$). It is determined by a titration of the water with a base (usually barium hydroxide) to an endpoint of 8·6 (Golterman *et al.*, 1978). Despite these reservations, however, there is still a fairly good linear relationship between pH and the logarithm of total acidity in acid mine drainage waters (Fig. 4.1); at pH values greater than 7 there is rarely very much acidity.

An understanding of the concept of total acidity is the key to understanding the differences between acid mine drainage and other acid ecosystems such as peat drainage and acid rain-affected areas where low pH values are coupled with low acidity. Table 4.2 gives a broad indication of the orders of some of the key variables in different acid waters. Peat drainage differs from the others as the source of its acidity is weak carboxylic acid groups rather than strong mineral acids; however whilst extreme acid mine drainage characteristically has a high concentration of total acid and high conductivity, where this is diluted by other streams in the catchment the water quality can come to resemble the soft, low-acidity waters affected by acid precipitation.

One of the most important effects of low pH is to destroy the bicarbonate buffer system, a feedback mechanism which controls the magnitude of shifts in pH. Below a pH of about 4·2 all carbonate and bicarbonate is converted to carbonic acid. This readily dissociates to water

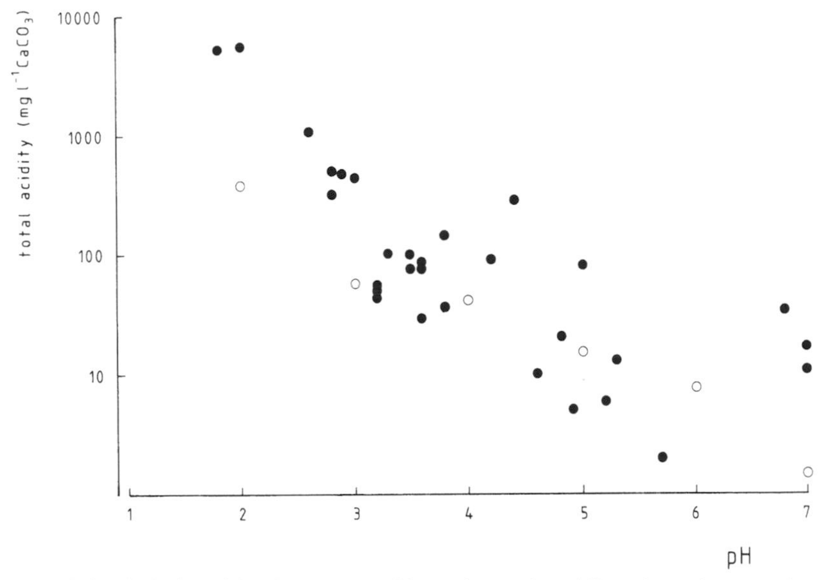

Fig. 4.1. Relationship between pH and total acidity, based on values measured in the field (closed circles) by Roback and Richardson (1969), Warner (1971), Koryak *et al.* (1972), Carrithers and Bulow (1973), and Sheath *et al.* (1982), and in the laboratory (open circles) by Bell (1971). Regression equation: $y = 3{\cdot}44 - 0{\cdot}39x$, $r^2 = 0{\cdot}68$; $p < 0{\cdot}001$.

Table 4.2
Comparison of waters receiving acid mine drainage, acid rain and peat drainage

Type of water	*Acid mine drainage*	*Acid precipitation*	*Peat drainage*
Source of acid	Oxidation of pyrites	Atmospheric sulphurous and nitrous oxides	Polyuronic acids in cell wall of *Sphagnum*
Main acids present	H_2SO_4	H_2SO_4 and HNO_3	—COOH groups in polysaccharides
pH	1·5–3·0	4–5·7	3·2–6
Acidity	110–64000	2·00–4·21	0·56–0·82
Conductivity	600–30000	< 50	< 100

Sources of values (ranges): acid mine drainage, Hargreaves *et al.* (1975); acidity in acid precipitation, Galloway *et al.* (1976); acidity in peat bogs, Clymo (1967); conductivity in peat drainage, Gorham (1956). Other ranges compiled from various sources. Acidity (as a measure of total concentration of strong acids) as mg litre^{-1} $CaCO_3$; conductivity (as a measure of total dissolved solids) as μS cm^{-1}.

and free carbon dioxide which may be lost to the atmosphere. There are two major effects of this: first, the water loses its capacity to buffer changes in pH and second, many photosynthetic organisms use bicarbonate as their inorganic carbon source. All aquatic organisms which live below pH 4·2 will need to be adapted to the lack of bicarbonate buffering, but aquatic plants, in addition, will need to be able to utilize free carbon dioxide as their inorganic carbon source for photosynthesis.

Once destroyed, the alkalinity of a water body may take some time to recover even if no further acid is added to the system. In one instance a stream took three and a half months to recover its bicarbonate buffering system following a pulse of acidity from a mine (Parsons, 1977). The rate of recovery was dependent upon the rate at which sources of alkalinity from the catchment could replenish the stream.

As flocculation of silt and clay is increased at low pH there may also be an increase in their rate of precipitation, resulting in water of low turbidity (Dills & Rogers, 1974), but this may be offset by the iron(III) precipitate formed when iron(II) salts in solution are neutralized (see below). Where both acidity and silt are present, the acidity has the greater effect on the biota and may to some extent mask the effect of the silt (Matter *et al.*, 1978).

A more general geochemical effect is to increase the rate of decomposition of clay minerals, feldspars and carbonates. Elements released by this include toxic metals, especially aluminium, but also silica and high concentrations of silica in acid mine drainage may stimulate the growth of acid-tolerant diatoms. To some extent the pH of high acidity water is buffered by aluminium silicates released by this mechanism. In experiments on the restoration of acid lakes King *et al.* (1974) showed that below pH 4·5 changes in pH were slow; however once the aluminium buffer system had been destroyed then neutralization was more rapid.

The effects of acid on the biota result in a general malaise (termed 'acidaemia' by Warner, 1971) which can completely kill some organisms. Specific physiological effects include an upset of the ionic balance across organism membranes and hydrolyzing or denaturing of cellular components, especially where these are exposed to the environment (Carlson-Gunnoe *et al.*, 1983). Calcium carbonate in the shells of molluscs and some crustaceans is likely to be dissolved.

Exposure to sources of neutralizing ions in the catchment can lead to reductions in the acidity of lakes over time. Brugam and Lusk (1986) used techniques developed by workers on lakes influenced by acid rain (reviewed by Battarbee, 1984) to calculate the pH of strip mine lakes in

Missouri from the diatom assemblages in sediment cores. Sampling of cores of lake sediments enables reconstructions of the pH of the lake throughout its history to be made. Out of twenty lakes, seven showed clear signs of having been acidic but had been neutralized over time; however most of the remainder appeared to have been alkaline when formed and had shown little sign of change. Neutralization, when it occurred, appeared to have been rapid, interpreted as a change from the aluminium buffering system in the acid lake to the bicarbonate buffering system. In one exception a lake which had originally been alkaline in its nature appeared to have become more acid with time, possibly because some previously buried sulphide minerals in the catchment had been re-exposed (Brugam & Lusk, 1986).

4.2.2 Ferric precipitate

The classic orange-hued flocs of iron(III) (ferric) precipitate ('ochre' or 'yellowboy') associated with acid mine drainage are composed predominantly of iron(III) hydroxide ($Fe(OH)_3$), resulting from the oxidation of pyrite. In reality the flocs are more complex and can contain a number of iron(III) oxyhydroxide hydrates such as $Fe(OH)_3(H_2O)_3$, various partially hydrolyzed forms (Barnes & Romberger, 1968; Clark & Crawshaw, 1979) and iron(III) hydroxysulphate complexes — $Fe(OH)(SO)_4)$ (Dugan, 1975). Some iron may also be precipitated as phosphate, effectively stripping this nutrient from solution. Other flocs may form from aluminium oxides and hydroxides (Parsons, 1968; Ashuckian & Finlayson, 1979) and, occasionally, other salts such as barium (Clark & Crawshaw, 1979) but all of these are almost white in colour.

These flocs form as the acid mine drainage becomes neutralized; at very low pH values the metal ions are soluble but as the pH rises some begin to precipitate out. The critical pH values are about 4·3 for iron(III) and 5.2 for aluminium (Skelly & Loy, 1973). Flocs have been noted at lower pH values but these probably formed when the pH was higher (Koryak *et al.*, 1972). As the pH of acid mine drainage is frequently below these values, the dissolved iron(III) salts will not precipitate out until a certain amount of neutralization has taken place. If conditions are favourable to oxidation of the pyrites and neutralization of the drainage this may be close to the outflow from the mine but in other cases it may not take place until much further downstream or where the acid stream joins a less acid river which can dilute the acidity and trigger the deposition of the floc. In Japan no iron(III) precipitate was found even 3 km below a mine discharging severe acid mine drainage to the River Aka (Noike *et al.*, 1983).

When the pH does rise the iron(III) salts come out of solution either to form colloids suspended in the water, fine suspended precipitates or heavier amorphous flocs. All of these can have severe effects on the biota. In suspension the floc reduces light penetration and so interferes with photosynthesis and the vision of consumers. It can also cause some physical abrasion (Eddlemon & Tolbert, 1983). When it settles out it can encrust rocks and stones, smothering all the benthic biota, filling gaps between stones and settling, especially in sluggish pools, to give a deep layer of enveloping deposit (Lackey, 1938; Hynes, 1960). Even after the source of acid mine drainage has been stopped, a severe spate can resuspend iron(III) deposits and, once again, affect biota for some distance downstream.

4.2.3 Heavy metals

Metal impurities may be found not only in wastes draining metal mines (see above) but also in coal seams where the acidity formed when pyrites are present can greatly enhance their solubility. A great deal of neutralization may be necessary to precipitate the metal ions as hydroxides (Table 4.3). Heavy metals which have been found at high concentrations in acid waters include nickel, copper, zinc and lead (Hargreaves *et al.*, 1975; Barton, 1978; Rasmussen & Sand-Jensen, 1979).

Metal ions are lost from solution by precipitation or adsorption, especially when the iron(III) and aluminium flocs are present. The extent of adsorption is dependent upon pH; in Cornish river samples receiving acid

Table 4.3
Minimum pH values for complete precipitation of metal ions as hydroxides or other salts[a]

Metal	*Minimum pH–hydroxides*	*Minimum pH–other salts*
Sn	4·2	
Fe(III)	4·3	
Al	5·2	
Pb(II)	6·3	6·0
Cu(II)	7·2	5·3
Zn	8·4	7·0
Ni	9·3	
Fe(II)	9·5	
Cd	9·7	
Mn(II)	10·6	

[a]Compiled from data in Down and Stocks (1977) and Eyres and Pugh-Thomas (1978).

mine drainage the ratio of bound to free copper and zinc increased as the pH rose (Johnson, 1986; Johnson & Thornton, 1987) and this can lead to increased concentrations of metals in the sediments as the pH rises (Lampkin & Sommerfield, 1986). This is a two-way process; a subsequent drop in pH can lead to the release of metals from the oxides (Tipping *et al.*, 1986). This is particularly relevant when a flush of more acid water flows down a stream, releasing metals which are bound this way in sediments, possibly with deleterious effects on the biota (Eyres & Pugh-Thomas, 1978).

The drop in pH is accompanied by a rise in the solubility of metals, making possible very high aqueous concentrations. Examples include 17·2 mg litre^{-1} copper and 114·4 mg litre^{-1} zinc at pH 3·6 in the Agrio River in south-west Spain (Cabrera *et al.*, 1984) and 190 mg litre^{-1} copper and 156 mg litre^{-1} zinc at pH 2·4 in West Squaw Creek in California (Filipek *et al.*, 1987). Accompanying this are changes in the partitioning of the metals with less adsorbed onto suspended particulate material and sediment (Johnson, 1986; Filipek *et al.*, 1987).

More complete coverage of the chemistry and biology of heavy metals is found elsewhere in this book (Chapters 3, 5, 6 and 7).

4.2.4 Turbidity

In addition to that associated with iron(III) deposits (Section 4.2.2) there are a number of other causes of turbidity associated with mine workings.

(i) Waste sedimentation ponds ('Tailings' ponds) are designed to trap the large amounts of inert rock particles washed out of mines and allow these to settle out; however these may occasionally overflow, with detrimental effects on the biota, especially if the waters had previously been kept clear of silt deposits.

(ii) Where acid drainage is treated with lime compounds to neutralize the acidity large volumes of calcium sulphate floc may be produced in addition to the iron(III) and aluminium hydroxides (Herricks & Cairns, 1977; Aanes, 1980).

(iii) Haul roads around mines, especially in rural areas, are often untarred and, consequently, very prone to erosion.

(iv) Suspended solids may be produced by spoil heaps long after the mine has been abandoned (Matter *et al.*, 1978) and these problems are likely to be more severe when the inhibitory effects of the acid and heavy metals prevent revegetation which can reduce erosion.

4.2.5 Other factors

Ferrous iron (iron(II)) may be present, especially when the drainage is incompletely oxidized. Its polluting effect is clear from its chemical oxidation:

$$4Fe^{2+} + O_2 + 10H_2O \rightarrow 4Fe(OH)_3 + 8H^+$$

Not only acid, but also iron(III) hydroxide is produced; and the process consumes oxygen. On the other hand, iron(II) hydroxide tends to remain in solution unless the pH is 9·5 or more. Consequently there is both a neutralization demand (200 mg litre^{-1} of Fe^{2+} can lower a water of neutral pH to pH 2), and a chemical oxygen demand upon the receiving waters (Barton, 1978). This can result in a drop in dissolved oxygen but it is rarely severe. The drop of 2 mg litre^{-1} noted by Letterman and Mitsch (1978) is typical of the demand on receiving streams. Whereas such a drop might affect the rich fauna of an unpolluted stream, the already restricted fauna of a stream polluted by acid mine drainage is unlikely to be reduced much more. Water issuing from underground mines is often cooler than ambient waters and this can offset some of the problems as oxygen is more soluble in water at lower temperatures.

An indirect source of oxygen demand may result from the reduced activity of the decomposer chains in the aquatic ecosystem. Allochthonous material washed into the stream will therefore not be broken down so rapidly but will tend to accumulate and be washed downstream. When this organic material finally reaches a zone in which the decomposer chain is less affected by acid mine drainage, the accumulated debris may give rise to a large biological oxygen demand (Koryak *et al.*, 1972).

As the acidity from acid mine drainage tends to dissolve minerals as it runs over them, it is usually associated with hard water. Sulphate, obviously, can reach very high concentrations — 10 000 mg litre^{-1} is not impossible — but other elements are also present in very high concentrations. The mean values of calcium from different studies includes 416 mg litre^{-1} (Barton, 1978), 800 mg litre^{-1} (Bosman, 1983) and 269 mg litre^{-1} (Hargreaves *et al.*, 1975). Some deleterious effects on the biota may occur because of this hard water but, in reality, such effects would be very difficult to separate from the other problems associated with acid mine drainage and no detailed information is available.

Chapter 5

Uptake and Accumulation of Heavy Metals

Accumulation of heavy metals by aquatic organisms provides an essential link between the concentrations of metals in the environment and the effect that these have on the biota. There is a more immediate interest in the related process, biomagnification, when heavy metals are passed through food chains which may end with man. The tragic instance of mercury pollution in Minamata Bay, Japan, is perhaps the best known example of this. At the same time three of the metals dealt with here, nickel, copper and zinc, cannot be considered solely as pollutants as they are essential micronutrients to plants and animals (copper and zinc) or animals alone (nickel). Thus, it is reasonable to assume, organisms are going to require the mechanisms to accumulate at least small quantities of these metals.

There is one concept which needs some explanation at the outset. This is the 'enrichment ratio' (Brooks & Rumsby, 1965) or 'concentration factor', the ratio of the concentration in the organism to the concentration in the water, as 'parts per million' ($\mu g\, g^{-1}$ and mg $litre^{-1}$ respectively) which is usually expressed as a dimensionless quantity. To be strictly dimensionless the concentration in the water should be expressed in the same terms as the concentration in the organism but as one litre of water weighs one kilogram the problem is partially circumvented and becomes one of convention alone. Nonetheless the enrichment ratio remains a useful tool for comparisons between metals.

What will become apparent in the rest of this chapter is how far there still is to go to understand thoroughly the processes of metal uptake. There are plenty of good studies documenting the total concentrations of heavy metals in organisms but there is still a long way to go before the level of understanding of the physiology of heavy metals approaches that of more important nutrients such as nitrogen, phosphorus and calcium.

5.1 PLANTS

Plants have one complication and one simplification over animals in studies of heavy metal uptake and accumulation. The complication is the cell wall which provides an additional barrier for metals to circumvent, whilst the simplification is their position at the base of the food chain. Amongst aquatic plants, however, there is a great number of growth strategies, each of which will present different opportunities for metal accumulation. Proximity to sediments, especially via roots in higher plants, is perhaps the most obvious of these. Because of this it is sensible to tackle the lower plants first. Even then it is not possible to disregard totally the sediments as a source of metals; some planktonic algae over-winter on the bottom muds (Denny & Welsh, 1979; Reynolds, 1984) whilst many attached forms are in almost continual contact with their sub-stratum.

Several of the basic features of metal accumulation by aquatic plants are illustrated in Figs 5.1–5.4 compiled from concentrations of metals in water and plants growing in the field. The criteria used to select the data were:

(i) that concentrations both in the water and plant were available in a tabular form;
(ii) that the plant grew naturally *in situ* and had neither been artificially transplanted nor had any artificial 'spike' been added to the water; and
(iii) the concentration in the plant was expressed on a dry weight basis.

In practice the dataset is limited partly by the practical difficulty of scanning all the available literature and partly because a large amount failed to satisfy all these criteria. Commonly the results were expressed in terms of the enrichment ratio rather than an absolute concentration and these were excluded although a figure could have been obtained by back-calculation. Linear relationships are usually found when concentrations in water and plants are expressed as Log_{10}, a technique which both satisfies the assumptions of normality made in regression analysis and allows data spanning several orders of magnitude to be presented clearly.

At least 43 species were represented (a few were identified only as genera and one was a composite 'phytoplankton' sample); these included re-presentatives of the Cyanophyceae, Rhodophyceae, Xanthophyceae,

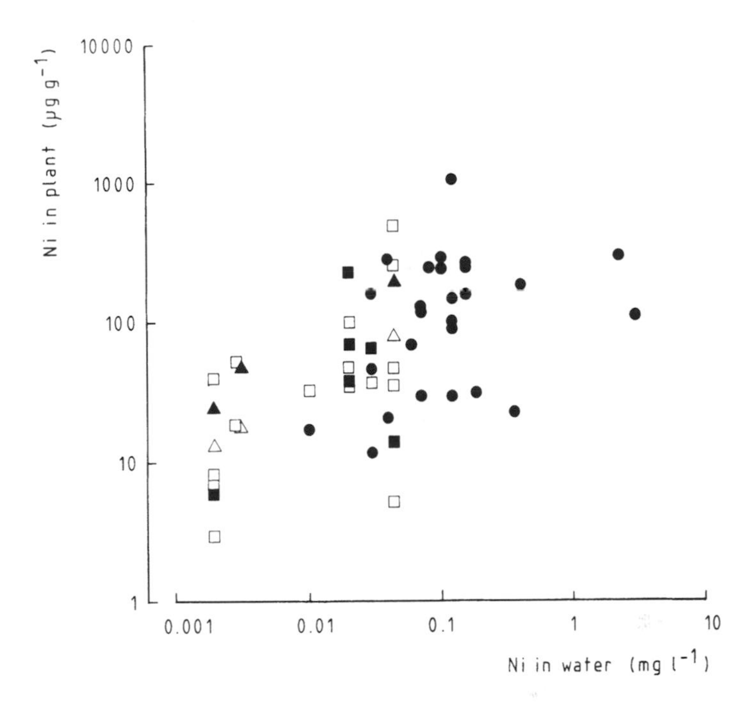

Fig. 5.1. Accumulation of nickel from water by algae (closed circles), pteridophytes (roots, closed triangles; shoots, open triangles) and angiosperms (roots, closed squares; shoots, open squares). Original figure compiled from published (Hutchinson *et al.*, 1975, 1976; Trollope & Evans, 1976) and unpublished (B.A. Whitton, personal communication) sources. See text for more details.

Chlorophyceae, Bryophyta, Pteridophyta and Sporophyta. Despite this variety there are still clear relationships for all four metals with a general trend for higher aqueous concentrations to result in higher concentrations in plants (Figs 5.1–5.4). Nonetheless there is still a considerable amount of scatter in all four graphs as well as some quite obvious differences between them. Sources of variation within graphs will include differences between species and in the environment of the sampling sites and, as these are compiled from values in the literature, differences in analytical techniques. Despite this almost two thirds of the variability for zinc (coefficient of determination, $r^2 = 0{\cdot}65$) is explained by the bivariate regression alone. Lower amounts of the variability for the other metals are explained in this way (Table 5.1).

The data for zinc span approximately one order of magnitude. Prominent among the outliers are two bryophytes well below the main trend

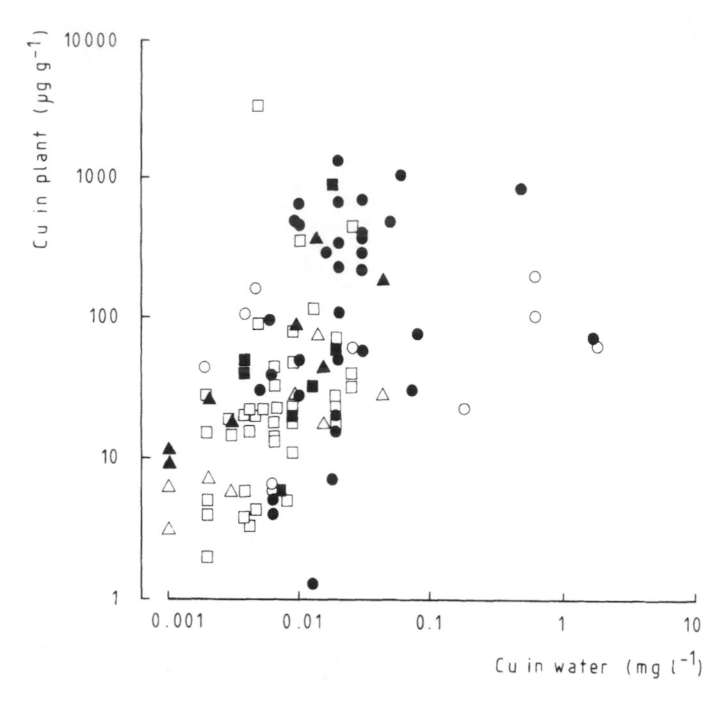

Fig. 5.2. Accumulation of copper from water by algae (closed circles), bryophytes (open circles), pteridophytes (roots, closed triangles; shoots, open triangles) and angiosperms (roots, closed squares; shoots, open squares). Original figure compiled from published (McLean & Jones, 1975; Hutchinson *et al.*, 1975, 1976; Trollope & Evans, 1976; Ernst & Marquenie-van der Werff, 1978; Cushing, 1979; Ray & White, 1979; Satake *et al.*, 1984; Fayed & Abd-El-Shafy, 1985) and unpublished (B.A. Whitton, personal communication) sources. See text for more details.

(bottom right); these are an aquatic liverwort *Jungermania vulcanicola* collected from acid streams in Japan (Satake *et al.*, 1984) where, presumably, the low pH limits the accumulation of metals. The two algae close to these are both *Hormidium rivulare* collected from high zinc sites, which may use exclusion as a tolerance mechanism. Increased secretion of mucilage has been observed around *Hormidium* subjected to high copper concentrations (Sorentino, 1985) which may help to exclude metals from the cell. The cluster of algae at the top right-hand corner were all collected from sites close to a zinc smelter with ambient concentrations up to 34·1 mg litre^{-1} zinc.

Zinc is a particularly good example to discuss as it is present in relatively high concentrations, easy to analyze, relatively soluble and a great number

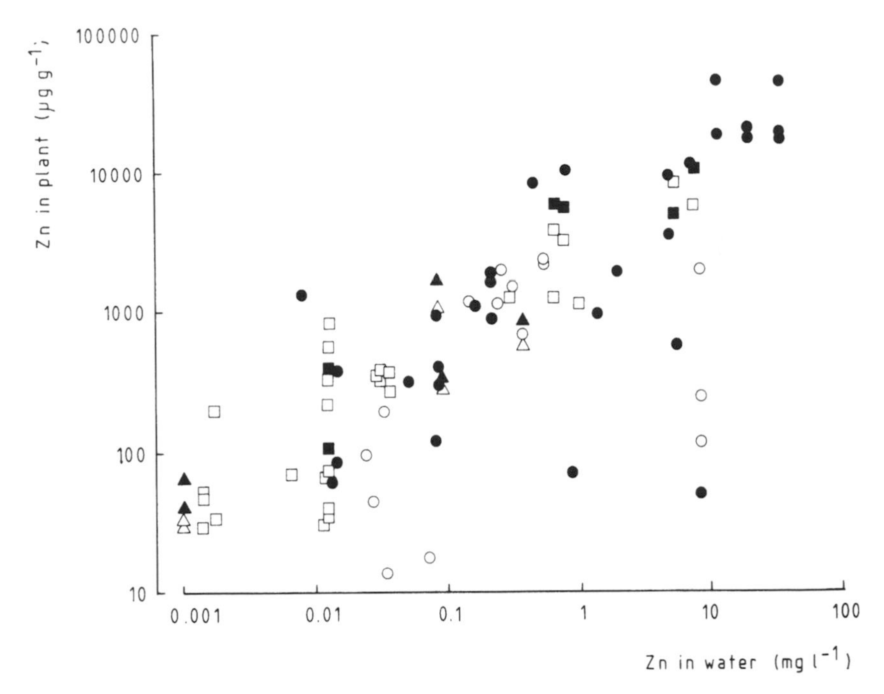

Fig. 5.3. Accumulation of zinc from water by algae (closed circles), bryophytes (open circles), pteridophytes (roots, closed triangles; shoots, open triangles) and angiosperms (roots, closed squares; shoots, open squares). Original figure compiled from published (Trollope & Evans, 1976; Cushing, 1979; Ray & White, 1979; Adams *et al.*, 1980; Say *et al.*, 1981; Satake *et al.*, 1984; Fayed & Abd-El-Shafy, 1985; Jones *et al.*, 1985) and unpublished (B.A. Whitton, personal communication) sources. See text for more details.

of studies have been concerned with it. All of the other metals fail to satisfy at least one of these criteria. However differences in the relationships can still be seen, both in their slopes and the absolute concentrations accumulated (= intercept; Table 5.1), indicating differences in the behaviour of the organisms with respect to the four metals.

High enrichment ratios are also apparent from these diagrams. Typically, these fall in the range 10^3–10^4 (e.g. Keeney *et al.*, 1976; Trollope & Evans, 1976), although these tend to decrease as the aqueous concentration increases (Figs 5.5–5.8), presumably representing the decreased availability of sites to bind metals as the concentrations increase. This is particularly marked for analyses of the algal samples collected from around a zinc smelter in South Wales (Trollope & Evans, 1976). Close to the smelter, where zinc concentrations were in excess of 10 mg litre^{-1} the

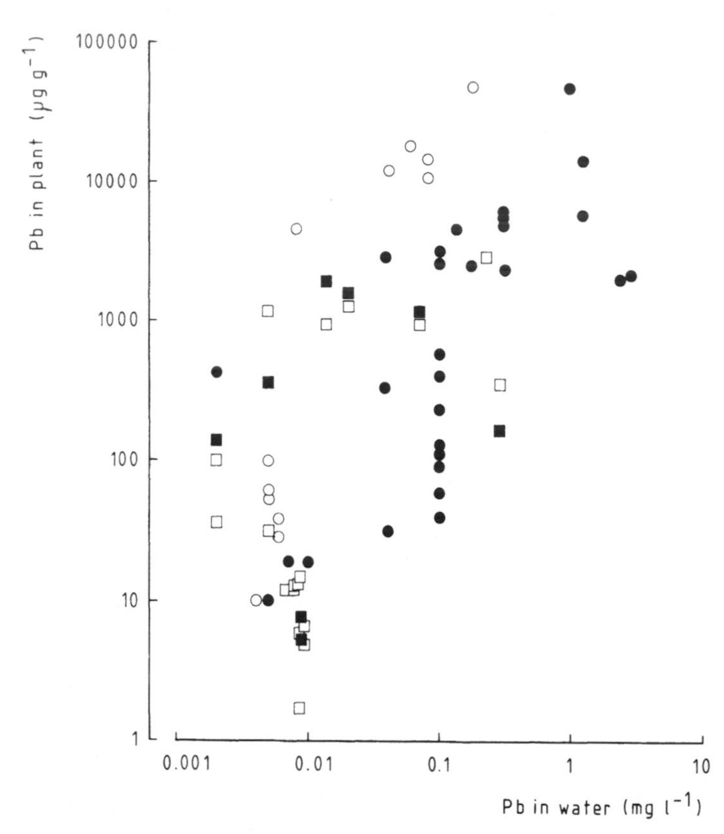

Fig. 5.4. Accumulation of lead from water by algae (closed circles), bryophytes (open circles) and angiosperms (roots, closed squares; shoots, open squares). Original figure compiled from published (McLean & Jones, 1975; Trollope & Evans, 1976; Welsh & Denny, 1976; Ray & White, 1979; Say *et al.*, 1981; Fayed & Abd-El-Shafy, 1985; Jones *et al.*, 1985) and unpublished (B.A. Whitton, personal communication) sources. See text for more details.

mean enrichment ratio was 1400; further away the ambient concentration dropped to less than 0·5 mg litre^{-1} whilst the mean enrichment ratio was 6700 (Trollope & Evans, 1976). Figures 5.5–5.8 also illustrate the large amount of variation in enrichment ratio. This occurs within as well as between species and reflects the influence of other environmental factors (Keeney *et al.*, 1976) and, perhaps, differential responses of the organisms to the elevated metal concentrations. The enrichment ratio of zinc in aquatic liverworts from Scotland was inversely correlated with pH (Caines *et al.*, 1985) and similar relationships could presumably be found for other factors which affect accumulation (see below). For zinc the range of

Table 5.1
Regression equations for $\log_{10}$ metals in plants plotted against $\log_{10}$ metals in the water from which the plants were removed, based on same data used to construct Figs 5.1–5.4

Equation	*n*	*F*	r^2
$\log_{10} Ni_{water} = 0{\cdot}382 \log_{10} Ni_{plant} + 2{\cdot}380$	45	14·25	0·249
$\log_{10} Cu_{water} = 0{\cdot}516 \log_{10} Cu_{water} + 2{\cdot}665$	80	17·19	0·181
$\log_{10} Zn_{water} = 0{\cdot}587 \log_{10} Zn_{water} + 3{\cdot}266$	75	136·95	0·652
$\log_{10} Pb_{water} = 0{\cdot}861 \log_{10} Pb_{water} + 3{\cdot}674$	65	38·74	0·381

n = number of samples; F = variance ratio; r^2 = coefficient of determination of regression. All relationships are significant at $p < 0{\cdot}0001$ (based on analysis of variance).

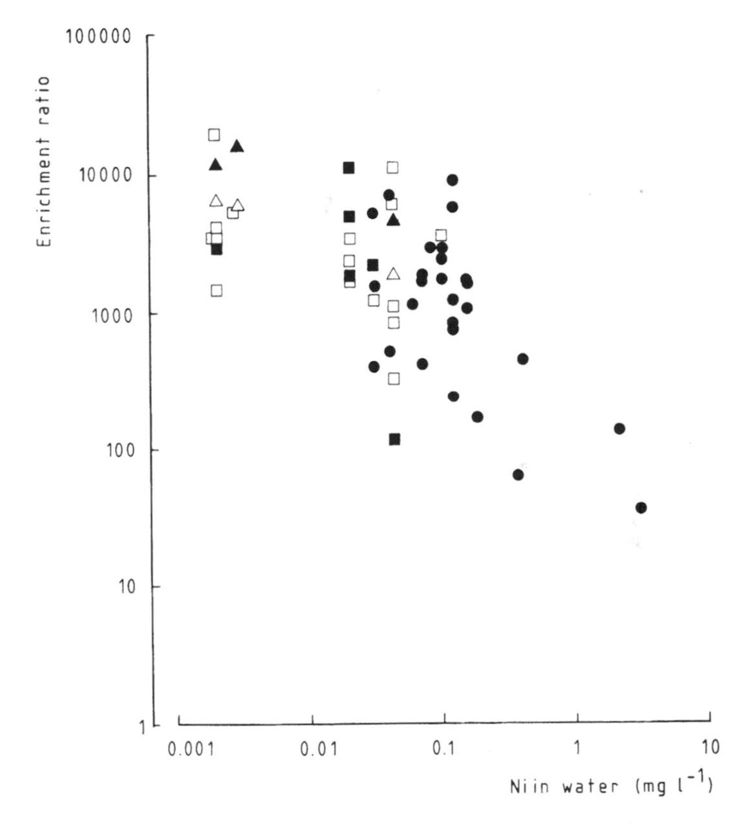

Fig. 5.5. Enrichment ratios (concentration in plant, in μg g^{-1}/concentration in water, in mg litre^{-1}) for nickel, based on same data as Fig. 5.1. See text for more details.

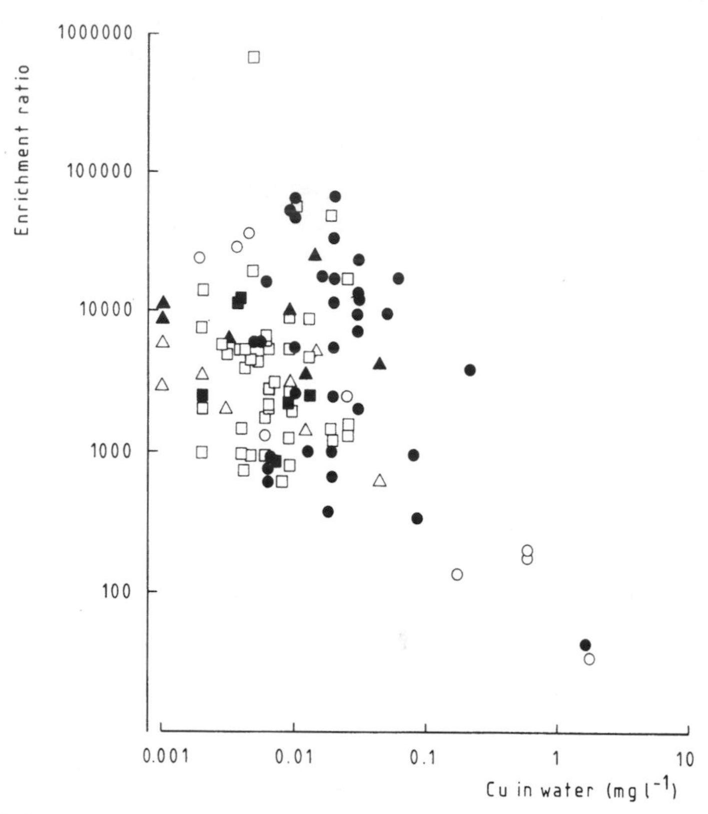

Fig. 5.6. Enrichment ratios (concentration in plant, in μg g^{-1}/concentration in water, in mg litre^{-1}) for copper, based on same data as Fig. 5.2. See text for more details.

enrichment ratios increases as the ambient concentration increases, perhaps indicating differential use of metal exclusion as a tolerance mechanism.

In one of the earliest studies Dietz (1973) compared enrichment ratios in aquatic bryophytes and angiosperms. His results clearly showed the highest ratios in the bryophytes, followed by two submerged angiosperms with divided leaves (*Ranunculus fluitans* and *Myriophyllum spicatum*) and the lowest ratios in emergent (*Sagittaria sagittifolia*) and floating-leaved (*Nuphar luteum*) angiosperm species. Few subsequent studies have been able to show these differences quite as clearly.

One further advantage of using enrichment ratios is that they allow comparison of accumulation of different metals although they may occur at different ambient concentrations; however when published values are

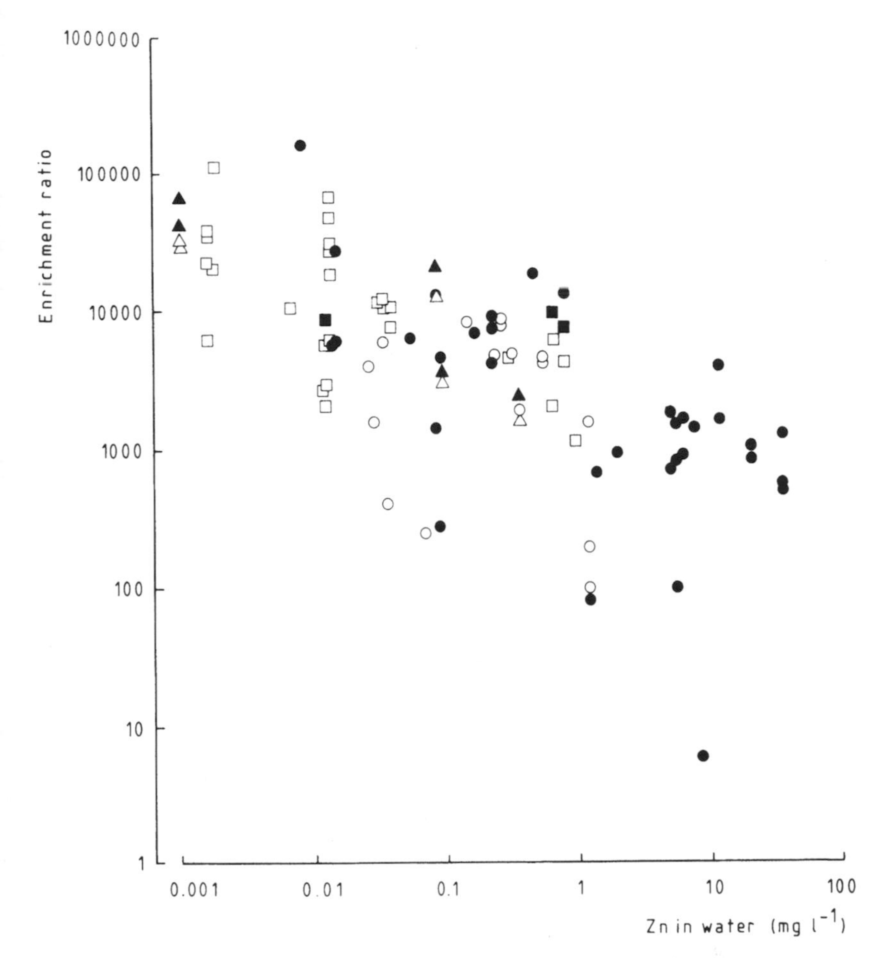

Fig. 5.7. Enrichment ratios (concentration in plant, μg g^{-1}/concentration in water, in mg litre^{-1}) for zinc, based on same data as Fig. 5.3. See text for more details.

compared (Table 5.2) no clear pattern emerges and, again, each case should be considered separately. However, in order to calculate enrichment ratios the concentration in both the water and the plant are required and most workers have preferred to deal directly with these.

Whitton and co-workers have made a number of intensive studies of individual species of algae and aquatic bryophytes. Metal concentrations in the red alga *Lemanea fluviatilis* collected from fast-flowing streams in Western Europe, many influenced by mining activities, were significantly correlated (on a $\log_{10}$–$\log_{10}$ basis) with aqueous concentrations of zinc,

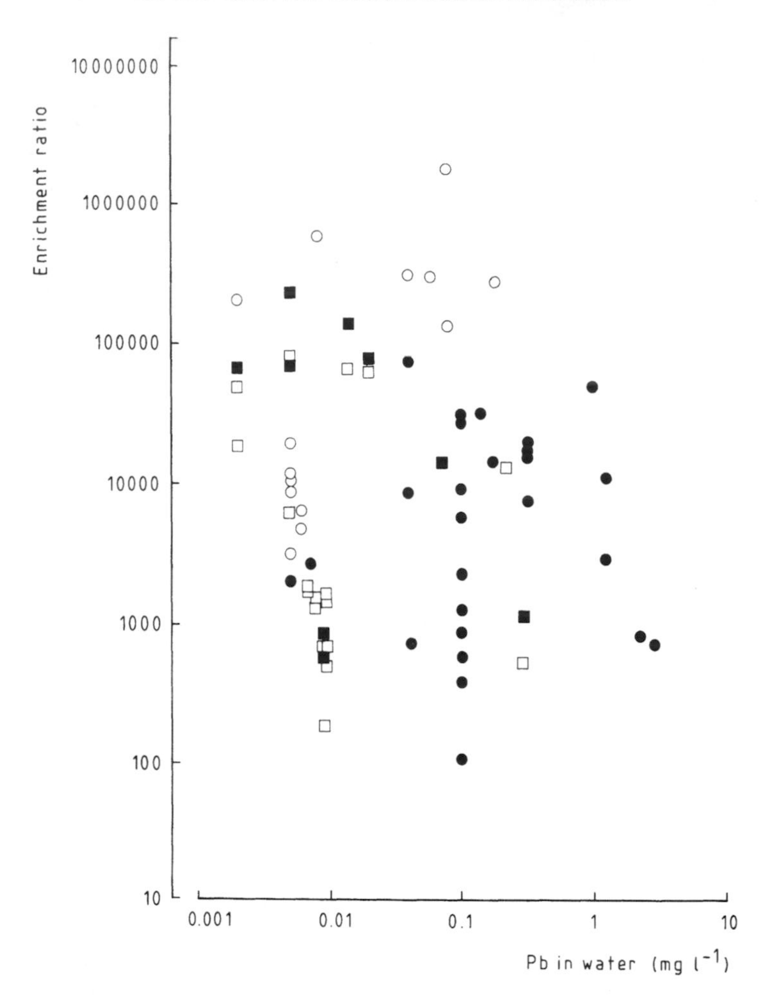

Fig. 5.8. Enrichment ratios (concentration in plant, in μg g^{-1}/concentration in water, in mg litre^{-1}) for lead, based on same data as Fig. 5.4. See text for more details.

cadmium and lead (Harding & Whitton, 1981). Similar results were found in subsequent studies on aquatic bryophytes (Whitton *et al.*, 1982; Say & Whitton, 1983; Wehr & Whitton, 1983*b*; Kelly & Whitton, 1988) although these also used multiple stepwise regression techniques in order to identify factors which affected accumulation and so produce improved equations. For the moss *Rhynchostegium riparioides* the bivariate equation accounted for 0·64 of the variation (as r^2) whilst the multivariate equation improved this to 0·84 (Wehr & Whitton, 1983*b*). Factors which were found to make

Table 5.2
Relative ranking of enrichment ratios reported in the literature

Study	*Species*	*Rank*
Algae		
Keeney *et al.* (1976)	*Cladophora glomerata*	Cu < Zn < Pb <Cd
Trollope & Evans (1976)	Various species	
	'Close' to zinc smelter	Zn < Ni < Pb < Cu
	'Near' zinc smelter	Zn < Ni < Cu < Pb
	'Distant' from zinc smelter	Ni < Pb < Cu < Zn
Bryophytes		
Dietz (1973)	*Fontinalis antipyretica*	Hg < Cu < Ni < Pb < Zn
	Hygroamblystegium	Ni < Zn < Cu < Pb
Wehr & Whitton (1983*b*)	*Rhynchostegium riparioides*	Ni < Ba < Co < Cu < Cd < Zn < Pb
Higher plants		
Dietz (1973)	*Ranunculus fluitans*	Hg < Ni < Cu < Pb < Zn
	Nuphar luteum	Hg < Cu <Pb < Zn < Ni
	Sagittaria sagittifolia	Ni < Cu
	Myriophyllum spicatum	Hg < Pb < Ni < Cu < Zn
Fayed & Abd-El-Shafy (1985)	*Eichhornia*	Cd < Pb < Cu < Zn
	Panicum	Cd < Pb < Cu < Zn
	Ceratophyllum	Cd < Pb < Cu < Zn

significant contributions to the variability included aqueous calcium and magnesium concentrations (increased concentrations decreased accumulation) and concentrations of manganese and iron in the plant (increased concentrations increased accumulation). These studies tended to be more successful for zinc, cadmium and lead than for other metals; however this largely reflects the initial choice of sampling sites and it is likely that similar relationships could be found for other metals as well. Altogether, datasets for seven species (three algae, four bryophytes) have been collected. For zinc, the aquatic bryophytes tended to accumulate higher absolute concentrations; however the green algae (*Cladophora glomerata* and *Stigeoclonium tenue*) had steeper slopes, indicating a greater change in the accumulated concentration for the same change in aqueous concentration compared with the bryophytes (Kelly & Whitton, 1988). The red alga *Lemanea* had a similar slope to the bryophytes but contained lower absolute concentrations.

Laboratory studies tended to corroborate some of these findings. Magnesium and calcium both decreased zinc accumulation by *Rhynchostegium*; aqueous manganese slightly decreased accumulation but manganese and iron in the plant had the reverse effect (Wehr *et al.*, 1987). They attributed this increase to extracellular oxides which trapped metals by adsorption and co-precipitation (see Chapter 3). These oxides are particularly apparent on the older, lower parts of the stems of aquatic bryophytes (Whitton *et al.*, 1982; Wehr *et al.*, 1987). Increased concentrations of aluminium and titanium, elements not normally associated with plant growth, were found along the thalli of the liverwort *Pellia endiviifolia* (Satake *et al.*, 1987). Increases in pH led to increased accumulation of metals in the field for the aquatic liverwort *Scapania undulata* (Whitton *et al.*, 1982) although this was less obvious for the other species studied, partly because of their narrower pH ranges. At very low pH values accumulation is much reduced. For example, the liverwort *Cephalozia bicuspidata* accumulated 1890 $\mu g\,g^{-1}$ and 1930 $\mu g\,g^{-1}$ from 27·8 and 26·9 mg $litre^{-1}$ zinc respectively at pH < 4 (Wehr & Whitton, 1983*a*); compare this with Fig. 5.3. The study of Satake *et al.* (1984) on *Jungermannia vulcanicola* from acid streams (pH 3–4·5) also shows metal concentrations well below those 'expected' from Fig. 5.3.

Amongst the algae, reduced accumulation at low pH values has been shown for *Chlamydomonas variabilis* (Harrison *et al.*, 1986) and *Chroococcus paris* (Les & Walker, 1984); however in another study *Chlamydomonas* accumulated the most nickel at pH 6·0, a feature attributed to the unusual glycoprotein structure of its cell wall (Wang & Wood, 1984). Differences have also been shown in the response of algae to different chemical forms of metals; a particularly interesting demonstration of this showed two organic compounds based on lead(IV) to be less readily taken up by *Ankistrodesmus falcatus* than inorganic lead(II) compounds (Wong *et al.*, 1987).

Sediments are rarely considered as sources of metals for algae and bryophytes. Experimental studies on *Rhynchostegium riparioides* (as *Platyhypnidium riparioides*) indicated that there was little direct uptake from the sediments (Hébrard & Foulquier, 1975) and field studies in the River Etherow, North-West England, showed no relationships between metal concentrations in plants and sediments for the mosses *R. riparioides*, *Fontinalis squamosa* or *F. antipyretica*, although pooled data did show a significant relationship (Say *et al.*, 1981). However, in a study such as this cross-correlations between concentrations in water and sediments and water and plants cannot be discounted.

For rooted aquatic pteridophytes and angiosperms the sediments cannot be excluded as a pathway for accumulation. The extent to which plants utilize minerals is dependent both on the species (Denny, 1972) and the metal (Welsh & Denny, 1979) but in most cases the concentration in the root is higher than that in the shoot (see Figs. 5.1–5.4). Essential minerals which are taken up by the roots are translocated to the shoots; in *Potamogeton crispus* and *P. pectinatus* copper accumulated from the sediments followed this pathway whilst lead was retained in the roots (Welsh & Denny, 1979). This partial exclusion of some metals from shoots is similar to the mechanism proposed for heavy metal tolerance in terrestrial plants (Bradshaw & McNeilly, 1981). Generally, submerged plants have the highest concentrations of metals, followed by floating leaved species and emergents. This has been shown both between (Heisey & Damman, 1982) and within (Aulio & Salin, 1982) genera. In copper- and nickel-polluted regions of Lake Ontario the highest concentrations were found in the totally submerged pteridophytes (Miller *et al.*, 1983).

Many workers have measured accumulation of heavy metals over time and these tend to show an initial period of rapid accumulation. In algae, bryophytes and angiosperms collected from Ullswater 20–70 $\mu g\, g^{-1}$ was accumulated in the first minute; this rate subsequently declined and after three hours was usually less than 10 $\mu g\, g^{-1}$ per minute (Everard & Denny, 1985*a*). Over the first 16 hours the rates of accumulation by living angiosperm shoots and shoots killed by immersion into 100% methanol were similar, supporting a generally-held hypothesis that uptake in the early stages is largely by passive adsorption (Pickering & Puia, 1969; Wehr *et al.*, 1987). Beyond this initial period the results become more ambiguous. Different workers have suggested active and passive mechanisms, based on manipulations of environmental conditions (light, temperature) and metabolic conditions (using inhibitors). Inhibitors of both respiratory and photosynthetic systems have been shown to reduce metal accumulation by aquatic plants (Pickering & Puia, 1969; Skowronski, 1984) but as plant cells expend energy maintaining electronegative conditions in the cytoplasm this may not be evidence of 'active uptake' *sensu stricto*. Moreover, evidence from manipulations of environmental conditions also must be approached with care as changes in metabolism may result in pH changes in poorly buffered media (Parry & Hayward, 1973). Low cytoplasmic concentrations of metals, coupled with the electronegativity of plant cells combine to make conditions generally favourable for 'passive' accumulation of heavy metals followed by removal into 'storage' areas. In blue-green algae accumulation of copper, zinc and lead into polyphosphate bodies has been demonstrated (Jensen *et al.*, 1986).

Most of these studies used analyses of whole plants or tissues; however analyses at the cellular and sub-cellular level are also important in order to understand differential toxicity and tolerance. A great many of the studies on all classes of plants emphasize the role of the cell wall. Although this is composed mainly of cellulose there are also subtantial amounts of other carbohydrates such as pectic materials which contain relatively high concentrations of carboxylic acids. These readily dissociate in water and the anionic carboxyl groups bind other cations in a form which is beyond the plasmalemma and so less toxic to the plant. High concentrations of metals in cell walls have been shown for algae (Silverberg, 1975), aquatic bryophytes (Burton & Peterson, 1979; Satake & Miyasaka, 1984; Wehr *et al.*, 1987) and angiosperms (Sharpe & Denny, 1976; Ernst & Marquenie-van der Werff, 1978); however it is also clear that significant concentrations may pass the plasmalemma into the cells. Detailed discussion of the fate of metals within cells is best left to Chapter 6 but an understanding of the basic distinction between intracellular metals which have crossed the plasmalemma into the cell and metals which are reversibly bound to the cell wall may help appreciate how environmental factors such as pH and calcium concentration affect accumulation and, perhaps, toxicity as well. In waters where calcium is the dominant cation on a molar basis then there will be competition between it and heavy metals for the exchange sites, the outcome of which will depend upon the relative affinities and concentrations of each. In a laboratory experiment on the aquatic bryophyte *Rhynchostegium riparioides* increasing concentrations of calcium in the medium resulted in decreased adsorption of zinc by the cell wall (Fig. 5.9; Kelly, 1986). However, inhibition of both intra- and extracellular cadmium accumulation by magnesium and calcium in the moss *Rhytidiadelphus squarrosus* was observed by Brown and Beckett (1985), suggesting that their effect was not confined solely to the cell wall. Similarly, experiments on *Chlamydomonas variabilis* showed a decrease in pH to reduce both extra- and intracellular uptake of zinc (Harrison *et al.*, 1986).

5.2 ANIMALS

Although some workers have suggested (e.g. Anderson, 1977) that the position in the food chain determines the degree of metal exposure in freshwater animals, several workers have disputed this and shown the carnivorous insects to contain lower concentrations than shredders and grazers (Burrows & Whitton, 1983; Smock, 1983*a*). As for plants, the sediments play an important role. In two river systems in North Carolina species which ingested much sediment (e.g. burrowing mayflies, Ephemeridae and some Chironomidae) had the highest concentrations of all the metals studied followed by filter feeders (e.g. Hydropsychidae), detritivores and algal grazers (Ephemeroptera, Plecoptera, Trichoptera) whilst the carnivores and surface feeding species such as Gerridae and Gyrinidae contained the lowest concentrations (Smock, 1983*a*).

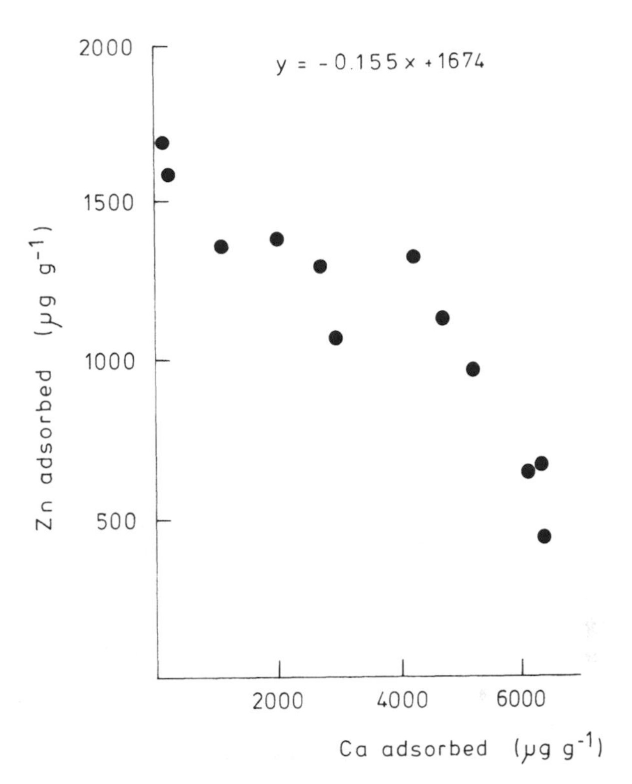

Fig. 5.9. Interactions between adsorbed calcium and zinc: effect of increasing concentrations of calcium in medium on subsequent zinc adsorption by cell walls of the aquatic bryophyte *Rhynchostegium riparioides*. Samples incubated for one hour at 15°C. The equation for the regression is: $y = 1674 - 0{\cdot}155x$, $r^2 = 0{\cdot}84$. From Kelly (1986).

Ephemeroptera (Mayflies) contained the highest and carnivorous species (especially the Plecopteran *Perla bipunctata*) the lowest concentrations of zinc, cadmium and lead in a stream polluted by mine waters in Northern England (Burrows & Whitton, 1983). In the River Hayle in Cornwall the highest concentrations of copper and zinc were found in 'free-living' Trichoptera (Caddis Fly) larvae (cased Trichoptera were thought to be protected), followed by Plecoptera, Odonata and Neuroptera larvae and adult Coleoptera (Brown, 1977*a*). High concentrations of copper in crustaceans have been attributed to its presence in the pigment haemocyanin (Anderson, 1977).

Nonetheless feeding habits may still have a significant effect on invertebrate metal concentrations. Smock's (1983*a*) study showed the gut contents to represent a major proportion of the whole-body concentration. Variations in the concentration in the gut were interpreted as reflecting feeding habits and the amounts of sediment ingested; hence the high concentrations in the burrowing mayflies. In Crane Fly larvae (*Tipula* spp.) two-thirds of the zinc content was associated with the gut (Ellwood *et al.*, 1976). Consequently many workers keep invertebrates alive but unfed for a period prior to analysis to allow purging of the gut contents. This may last two (Manly & George, 1977), three (Ellwood *et al.*, 1976) or four (Ireland, 1975) days.

Only a few correlations between concentrations in animals and water have been reported. For example Brown (1977*a*) found a correlation only between the free-living Trichoptera and copper and zinc, Burrows and Whitton (1983) found one between three Plecopterans and a Coleopteran and zinc. However the size of the datasets used in each case were much smaller than those used in studies on plants and more extensive sampling of particular groups may reveal better correlations. Alternatively some invertebrate species may show better correlations between their individual tissue concentrations and substrate concentrations. This was shown for the crustacean *Asellus aquaticus* (for lead and copper) and the leech *Erpobdella octoculata* (for copper) in the River Irwell, North-West England (Eyres & Pugh-Thomas, 1978). In this study the water concentrations were lower than expected from the impoverished fauna and it was thought that the high pH leads to rapid removal of the metals to the sediments which acted as the main source of contamination.

The second major pathway in invertebrates is accumulation from the water itself. Rainbow and Moore (1986) and Smock (1983*b*) observed exponential decreases in concentration with increasing organism size, a relationship which was not affected by clearing the gut or the concentration of available metal in a more detailed study of the ephemeropteran *Stenonema modestum* (Smock, 1983*b*). This he interpreted as indicating adsorption onto the exoskeleton. In all, the gut represented 52% of the chromium content of *S. modestum*, the exoskeleton, 33% and the remaining 15% was assumed to be adsorbed internally. The exoskeleton does not represent such a large proportion in all invertebrates; in the crustacean *Daphnia magna* adsorption represented only a small proportion of the total nickel content and this was periodically reduced further by ecdysis (Hall, 1982). In another crustacean, two populations of the isopod *Asellus meridianus* populations collected from polluted ('tolerant') and clean ('non-tolerant') stretches of the River Hayle in Cornwall showed different behaviour. Both accumulated copper and lead from experimental solutions at similar rates but only the tolerant populations were able to accumulate metals from 'food' (soaked elm leaves, *Ulmus procera*), the non-tolerant populations tending to die. Extensive histochemical and ultrastructural examination identified the hepatopancreas as a site for copper storage in both populations but the deposits in the hepatopancreas of the tolerant population were associated with sulphur, possibly in the form of sulphydryl group (Brown, 1977*b*).

Concentrations of several metals higher than those in the sediment were found in the detritivorous and phytophagous snail *Physa* sp. (Enk & Mathis, 1977) but not in filter-feeding freshwater clams (Mathis & Cummings, 1973) which may, again, indicate the importance of the feeding method in determining metal concentrations in invertebrates. However this is only a partial explanation and, in a study of two lakes in New Jersey, concentrations of lead in grazing snails were more strongly correlated with aqueous concentrations than the concentration in the 'aufwuchs' (epiphytic algal communities), macrophytes or in various sediment fractions (Newman & McIntosh, 1982). Indeed the lowest lead concentrations were measured in a burrowing species, *Campelona decisum*, despite its close association with the sediments.

Tissue localization has been studied in the bivalve *Anodonta anatina* in the River Thames (Manly & George, 1977). Here, the softer parts (mantle, ctenida, kidneys) contained the highest concentrations of metals whilst the adductor muscles and the foot contained generally lower concentrations. In the freshwater snail *Taphius glabratus* the liver contained the highest concentrations of zinc, cadmium and lead accumulated from experimental solutions (Yager & Harry, 1964). The shell of snails can also accumulate certain metals; in phytophagous species in Ullswater lead was found to be bound inexchangeably into the structural material of the shell but concentrations here were several times lower than in the foot and the digestive gland (Everard & Denny, 1984). Nonetheless, when snails from contaminated sites were transferred to lead-free conditions their total lead concentration returned to near background levels. It is probable that localization in molluscs may vary depending upon the stage of the life-cycle, growth rate, food needs and other factors (Weatherley *et al*., 1980) which make generalizations very difficult.

There has been a lot of interest in heavy metal concentrations in freshwater fish, especially those which are eaten regularly. Many of the fish which are eaten by man are carnivorous; however, as for the invertebrates, there are not necessarily higher concentrations in these groups. In a contaminated eutrophic lake in Michigan the fish contained lower concentrations of lead than macrophytes, zooplankton or sediment (Mathis & Kevern, 1975). Tissue localization studies generally show that the highest concentrations of heavy metals are found in the liver (Abo-Rady, 1979; Bendell-Young *et al.*, 1986) along with the gills (Solbé & Cooper, 1976; Norris & Lake, 1984) and the kidneys (Cowx, 1982) whilst concentrations in the muscle are amongst the lowest (Merlini & Pozzi, 1977; Murphy *et al.*, 1978). These are unlikely to represent a major human health hazard unless fish constitute a major part of the diet. Use of multiple regression techniques on analyses of miller's thumb *Cottus gobio* showed a close relationship between the concentrations of metals in organs and in the whole body (Moriarty *et al.*, 1984). They concluded that no additional benefit was conferred by using organ metal concentrations unless the weight of the tissue was also available. When transferred out of a metal-enriched environment copper was readily lost from all body compartments of the Stone Loach *Noemacheilus barbatulus* except the liver and the vertebrae (Solbé & Cooper, 1976) and it is possible that in the rainbow trout *Salmo gairdneri* it may be bound to a metallothionein, a metal-binding protein (Laurén & McDonald, 1987*b*). As for the invertebrates there are not necessarily greater tissue concentrations at higher trophic levels. Perch (*Perca fluviatilis*) and pike (*Esox lucius*) from sites in north-west England had significantly lower concentrations of zinc, cadmium and lead than roach (*Rutilus rutilus*) and bream (*Abramis brama*) (Badsha & Goldspink, 1982). However, when food species susceptible to heavy metals are eliminated, the diet of predators may shift towards tolerant species, which may either exclude the metals or accumulate and store them. The latter strategy may serve to enhance the transfer of metals through the food chain (Dallinger *et al.*, 1987).

With organisms as mobile as fish it has been difficult to establish good relationships between environmental and tissue concentrations of metals; however approximate relationships have been found for lead concentrations in three species of trout in western USA (Pagenkopf & Neuman, 1974) and a general enrichment of metals has been observed in the brook trout *Salvelinus fontinalis* in the River Leine below Gottingen, West Germany (Abo-Rady, 1979) and for the perch *Perca fluviatilis* between metal-enriched and control lakes in Switzerland (Hegi & Geiger, 1979). In a survey of fourteen lakes in Ontario cadmium and mercury concentrations in fish were correlated with sediment concentrations whilst copper and zinc concentrations were relatively constant between species (Johnson, 1987). A better (though still quite approximate) relationship was found between ambient zinc concentrations and concentrations in the scales of Atlantic salmon *Salmo salar* and brown trout, *S. trutta* from sites in North Wales (Abdullah *et al.*, 1976). On a $\log_{10}$–$\log_{10}$ basis there was evidence of a minimum concentration of zinc (*ca* 400 $\mu g\ g^{-1}$ at aqueous concentrations up to 0·02 mg $litre^{-1}$) in the scales of *Salmo trutta* above which it was accumulated linearly. The results for *S. salar* were harder to interpret, presumably because of its migratory lifestyle, but the same general pattern was apparent. In a study of metal concentrations in arctic char (*Salvelinus alpinus*) interpretation was also confused by the presence of environmentally distinct populations in adjacent lakes, each with different food preferences (Cowx, 1982).

5.3 BIOLOGICAL MONITORING USING METAL ACCUMULATION

Data on metal accumulation by aquatic organisms clearly can be put to a number of uses including inferring the sources and fates of heavy metals and calculating the risks to human health from eating contaminated fish and invertebrates. A number of workers (mainly biologists) have also suggested that plants and animals may have a more direct role to play in monitoring water quality, complementing and in some cases replacing conventional water analysis. Although several of the studies discussed above were ostensibly 'monitoring' the environment, there has to date been little attempt to develop routine methods that water monitoring bodies can put to practical use. Clearly there ought to be a concentration on a few of the more effective methods.

Butler *et al.* (1971) and Phillips (1979) have tried to summarize the features of marine organisms which make good monitors; some of these apply equally well to freshwater organisms:

(i) the organism should be able to accumulate the pollutant to relatively high concentrations without being killed by the levels encountered;

(ii) it should be sedentary, in order to be representative of the area of collection;

(iii) it should be both abundant, well-known taxonomically and relatively easy to identify;

(iv) it should be available throughout the year; and

(v) it should be of a reasonable size so that adequate tissue will be available.

Finally, one extra point:

(vi) its general biology and ecology should be well understood.

Of all the different groups the filamentous algae, the macrophytes (*sensu lato*) and the molluscs are probably the most promising (Stokes, 1979). Points in favour of plants are (Whitton *et al.*, 1981):

(i) dried plant materials are easier to store and transport than water samples;
(ii) plant samples give an 'integrated' record of pollution within a particular system;
(iii) plants can be harvested after a pulse of contaminated water has passed downstream, so it is possible to pinpoint pollution after it has happened; and
(iv) data may be easier to interpret than similar data from animals which may be complicated by differing diets and avoidance behaviour.

Two of these points required additional comment. The concept of an 'integrated' record of pollution (ii) can cause confusion as the plants do not 'integrate' (*sensu stricto*) metal concentrations over time. Where workers have used this they have been trying to convey the impression that plants 'smooth out' the infinitesimal small changes in metal concentrations over time, in the manner (but not the mechanism) of a 'moving average'. This is particularly important under conditions where ambient conditions are constantly changing. On the other hand some plants may also be used to detect isolated incidences of pollution after they have occurred (iii). The aquatic bryophytes in particular have been identified for this purpose by virtue of their rapid rates of accumulation, but relatively slow rates of loss (Wehr *et al.*, 1987). Analyses of tissue concentrations of chromium in *Rhynchostegium riparioides* was used to identify the source of an intermittent input to the River Etherow in north-west England (Say *et al.*, 1981) and analyses of *Fontinalis antipyretica* identified an intermittent zinc effluent at a time when no discharge was taking place (Say & Whitton, 1983; Fig. 5.10). Subsequent studies showed that the extent to which *Rhynchostegium riparioides* retained metal was partly dependent upon the length of the initial exposure (Kelly *et al.*, 1987); however this may be a valuable technique for identifying some intermittent sources.

No one species is likely to be available at all sites and to overcome this a package of ten plants has been suggested which, it was hoped, would cover most eventualities within northern Europe (Whitton *et al.*, 1981). Although a zoologist may wish to add a couple of common molluscs to this list it remains a useful starting point and fulfils most of the criteria that a pragmatic monitoring biologist would set. Unfortunately, recent EEC

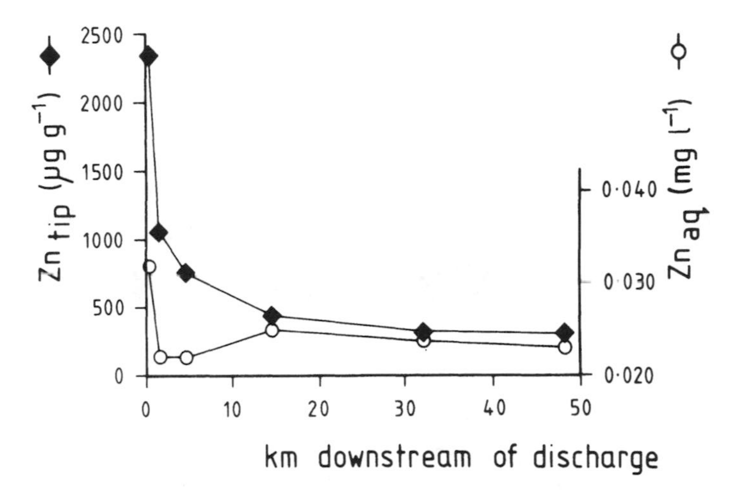

Fig. 5.10. Changes in zinc in water and 2-cm apical tips of the aquatic bryophyte *Fontinalis antipyretica* downstream of an intermittent discharge entering the River Wear, northern England, but sampled at a time when no pollution was entering the river. (Reproduced from Say & Whitton (1983), *Hydrobiologia*, **100**, 245–60 with permission from Dr W. Junk (Publishers).)

legislation on cadmium, along with mercury, the first of the heavy metals to receive such treatment, outlined biological criteria in terms of tissue concentrations in a characteristic mollusc. Realistically, however, there is almost certainly room for both points of view.

Chapter 6

Toxicity and Tolerance to Heavy Metals. I. Plants

Surprisingly, relatively few workers have critically assessed the chemical features which define a metal's response within an organism. In an attempt to replace the (in their opinion) ambiguous term 'heavy metal' with a more biologically relevant system Nieboer and Richardson (1980) resurrected an older system of classification of metal ions into three groups based upon their affinity for different ligands. These were: Class A, which tended to seek oxygen-containing ligands and comprised all the macronutrient metals such as potassium and calcium and no heavy metals except barium; Class B, nitrogen and sulphur-seeking metals, and 'borderline' metals, which had intermediate properties. Class A was sharply distinct from the borderline group whilst there was a quite fuzzy distinction between the borderline group, which included most of the common heavy metals, and Class B with copper(I), silver, gold and mercury.

This gave Nieboer and Richardson (1980) a set of criteria with which the actions of different heavy metals could be compared. For example the Class B metals, which are notoriously toxic, may bind tightly to sulphydryl groups in amino acids such as cysteine or to nitrogen centres in amino acids such as lysine and so disrupt the catalytic activation centres of enzymes. They may also displace endogenous borderline ions such as zinc or copper from enzymes or form lipid-soluble organometallic complexes (see Chapter 3) which can readily penetrate biological membranes. The toxicity of a borderline metal will vary depending upon the amount of Class B features which it displays but it will be able to displace Class A metals and, depending upon their relative affinities, other borderline metals. An example that Nieboer and Richardson (1980) gave from their own work was the replacement of nickel for zinc in the respiratory enzyme carbonic anhydrase. Based on both chemical (affinity for sulphide) and biological data our four heavy metals are ranked (in order of increasing toxicity): zinc (some Class A properties) < nickel < copper < lead (borderline — Class B).

Although such chemical classifications are valuable they only go part of the way to 'explain' the effects of heavy metals. An alternative, more 'bio-centric' standpoint, examined the metal as the organism 'saw' it (Stokes, 1983); here the metal speciation, the influence of other ions, the inherent tolerance of the cell and the physiological condition of the cell were all considered as modifying the organisms' response to the metal. Other workers go further; what the metal is, they say, is of secondary importance. Rather it is the degree of metal pollution *per se* rather than the particular metal which determines the (biological) species present (Foster, 1982*a*).

6.1 FIELD STUDIES

Amongst the earliest observations of the effects of heavy metals on aquatic communities are records of the reduction of both flora and fauna in the lead and zinc polluted rivers of West Wales (Carpenter, 1924). Immediately below the mines the flora was restricted to some growths of the red algae *Batrachospermum* and *Lemanea* and some mosses and liverworts. Recoveries from the pollution were observed both downstream of the mine and over time in reaches close to the mine after the mining had ceased. A general reduction in species due to heavy metal pollution has been a theme of a number of other studies although, for practical reasons, spatial studies are more common. For example a negative correlation between species numbers and the logarithm of the dissolved zinc concentration was observed in a study of a small catchment draining an abandoned lead mine in the Northern Pennines (Say & Whitton, 1980). Although the most polluted reach contained 25·6 mg litre^{-1} zinc and 2·06 mg litre^{-1} lead, 25 species, all algae, were recorded from it. Another study of communities, this time in artificial recirculating streams, showed a geometric decrease in the number of species colonizing glass slides as the zinc concentration increased (Williams & Mount, 1965). *Cladophora glomerata* was the dominant species in the control channel and several other species were epiphytic upon it. Whitton (1980) speculated that the decrease in species numbers at the polluted sites was due partly to the absence of *Cladophora* and its epiphytes. However, amalgamation of a mass of data of this nature indicated that the decrease in species numbers as zinc concentrations increased was a widespread phenomenon (Whitton & Diaz, 1980).

Conditions at the higher concentrations of metals represent extreme environments (*sensu* Brock, 1969) with the organisms undergoing considerable environmental stress. Those few species which can tolerate the conditions not only survive, they may well thrive in the reduced competition. Records of visually obvious growths of algae at such sites include the blue-green alga *Plectonema* on seepages draining tailings piles in the Old Lead Belt, Missouri (> 21 mg litre^{-1} zinc; Whitton *et al.*, 1981a), the green algae *Mougeotia* sp. (> 8·65 mg litre^{-1} zinc; Patterson & Whitton, 1981; Fig. 6.1) and *Stigeoclonium tenue* (> 1·31 mg litre^{-1} zinc; Armitage, 1979) in the Northern Pennines, Great Britain and *Hormidium rivulare* in the French Pyrénèes (> 42·5 mg litre^{-1} zinc; Say & Whitton, 1982). The reasons for these massive growths are not clear, nor is it fully understood why other sites with otherwise very similar water chemistries lack them. Massive growths of *Chlorella*, *Scenedesmus* and *Chlamydomonas* in dialysis bags placed in a copper-polluted river in Connecticut have been attributed to a lack of invertebrate grazers (Klotz, 1981), either less

Fig. 6.1. Caplecleugh Low Level, an adit draining a disused lead mine in the Northern Pennines, England. Throughout much of the year the short stream draining the adit into the River Nent is dominated by luxuriant growths of the green alga *Mougeotia* (light-coloured areas).

tolerant than the plants to heavy metals or unable to cope with the algal growths through a smothering of the substrate (Armitage, 1979), although deoxygenation of the water at night by algal respiration may also play a part. Such opportunistic growth also occurs when competing algal species are eliminated by heavy metals. In an oligotrophic stream in California, artificially dosed with copper one co-dominant species, the blue-green alga *Lyngbya* sp. was eliminated and replaced by the other co-dominant species, the diatom *Achnanthes minutissima*, the standing crop showing little change (Leland & Carter, 1984).

The total number of species alone is not necessarily an indication of the presence or absence of metal contamination and the logical step on from here is to make more detailed analyses of the composition of communities. In a survey of 190 stream sites the green algae *Ulothrix zonata* and *Microspora amoena* were shown to be susceptible to zinc and were absent at concentrations greater than 0·1 mg litre^{-1} whilst *Hormidium rivulare* and *Mougeotia* sp. were found at zinc concentrations greater than 10 mg litre^{-1} (Say & Whitton, 1981). These properties often appear to be at the specific rather than the generic level; the genera *Ulothrix* and *Microspora* were both represented by species with no resistance and with high resistance to zinc. Unfortunately there do not appear to be any algal indicators of heavy metals in the way that *Euglena mutabilis* often is of highly acidic waters (see Chapter 8) or that the so-called 'copper moss' *Mielichhoferia elongata* is said to be of terrestrial copper deposits (Wilkins, 1977).

Data of this type become more useful when combinations of species with otherwise similar environmental preferences are considered. For instance, it is well known that the green algae *Stigeoclonium tenue* and *Cladophora glomerata* are both common in eutrophic base-rich waters but *C. glomerata* is far less tolerant of heavy metals. The presence of *Stigeoclonium* in the absence of *Cladophora* is one indication that the water may contain heavy metals (Whitton, 1984). Analysis of algal communities of the copper-, zinc- and lead-polluted sites on the Rivers Hayle and Gannel in Cornwall showed no species which 'indicated' heavy metal pollution although a *Microspora*-dominated association comprising *M. stagnorum*, *M. pachyderma*, *M. willeana* and *Mougeotia* cf. *parvicula* could act as an 'indicator' community (Foster, 1982*a*). Several of the species she observed at metal-polluted sites were slime producers (e.g. *Batrachospermum*, *Zygogonium*, *Hormidium*).

The tolerance of the dominant species of diatom at a site, coupled with the absence of less tolerant forms, has been suggested as one guide to the extent of heavy metal pollution (Besch *et al.*, 1972); however when results from a number of studies are compared no clear picture emerges. Different studies have shown the diatom *Synedra ulna* both to be highly tolerant (Williams & Mount, 1965; Besch *et al.*, 1972; Austin & Munteanu, 1984) and highly sensitive (Say & Whitton, 1980) to zinc pollution. In streams draining copper, zinc and lead mines on Vancouver Island, British Columbia there were reductions in both the number and diversity of species below the mine but the composition of the communities which colonized glass slides was quite different to the previous studies. Diatoms dominated above and below the mine but green algae, which were abundant above the mine during the summer, were virtually absent below (Deniseger *et al.*, 1986). The numerically most abundant taxa above and below the mine included the diatom *Achnanthes minutissima*, recorded as 'resistant' by Say and Whitton (1981) but the absence of green algae, particularly *Mougeotia* which was abundant above the mine, is quite different to observations made by other workers. Stokes (1983), for instance, regards the green algae as more tolerant than either the blue-green algae or the diatoms; although Whitton and Diaz (1980) observed that generally increasing zinc concentrations had similar effects on all of the major phyla.

There are fewer field observations on the effects of heavy metals on aquatic macrophyte communities. In acidic soft-water lakes in Ontario (many in the Sudbury area) there was a reduction in the species diversity of angiosperms with decreasing distance from the smelter and as the concentration of sulphate in the water increased (Gorham & Gordon, 1963). They realized that the sulphate itself was probably not the toxic agent and suggested instead that nickel and copper may be responsible. This was later confirmed (Yan *et al.*, 1985); in this study a relationship between heavy metals and species numbers was shown for angiosperms but not for the charophytes nor the bryophytes. The most tolerant species were the bryophyte *Amblystegium riparium* (as *Leptodictyum riparium*) and *Eleocharis acicularis* var. *submersa* whilst *Utricularia vulgaris* and *Potamogeton epihydrus* var. *nuttallii* were the least (Gorham & Gordon, 1963). The effects of copper and zinc on river plants was also studied in another region of Canada, the Northwest Miramachi river system in New Brunswick (Besch & Roberts-Pichette, 1970). Alongside the river there were belts of gravel which were periodically inundated; at the sites closest to the mine where the river water could contain up to 12·1 mg litre^{-1} copper and 65·5 mg litre^{-1} zinc these were barren and as the metal concentrations decreased so the plant cover increased. In regions with medium to high pollution there was a sparse cover of *Equisetum arvense*,

Cyperaceae and Graminaceae and further downstream, where the pollution was lower, dicotyledons returned and only the submerged vascular plants were eliminated. Monocotyledons have been observed growing in or close to water containing high concentrations of heavy metals in a number of studies in Australia (*Phragmites*, *Typha*, *Juncus*; Weatherley *et al.*, 1967), North America (*Eleocharis*, *Scirpus*, *Typha*; Whitton *et al.*, 1981*b*) and Europe (*Phragmites*, *Typha*; Whitton, 1980).

6.2 LABORATORY STUDIES

Like terrestrial plants (Bradshaw & McNeilly, 1981) very few aquatic plants are known to be naturally tolerant of heavy metals; *Typha latifolia* may be one of the few exceptions (McNaughton *et al.*, 1974; Taylor & Crowder, 1984). For most plants some measure of genetic adaptation is required and for this the algae, with their short generation times, are ideally suited. Populations of the green algae *Scenedesmus* and *Chlorella* isolated from lakes close to the Sudbury smelter in Ontario were tolerant to elevated concentrations of nickel and copper compared to other isolates obtained from culture collections (Stokes *et al.*, 1973); however by subculturing the non-tolerant isolates into successively higher concentrations of nickel and copper a degree of tolerance was induced (Stokes, 1975*a*). After eight generations there was a doubling of the tolerance limits of both species to the two metals. A similar feat was performed on the blue-green alga '*Anacystis nidulans*' (≡ *Synechococcus* sp.; Rippka *et al.*, 1979) for zinc over 75 subcultures (Shehata & Whitton, 1982), raising the concentration which was strongly inhibitory to growth (i.e. that which just permits detectable growth) from about 2 to 15 mg litre^{-1} zinc. Resistance was rapidly regained even after the tolerant strain was subcultured twenty times in the absence of zinc. This, they suggested, along with similar results over 25 subcultures for cobalt, nickel, copper and cadmium (Whitton & Shehata, 1982), indicated that the resistance was due to the selection of spontaneous mutants. Tolerance to one metal did not necessarily confer tolerance to other metals in this study but cadmium-resistant *Nostoc calcicola* were able to tolerate elevated concentrations of zinc and mercury along with the antibiotics neomycin and chloramphenicol (Singh & Pandey, 1982). In contrast to these studies demonstrating the evolution of resistance over several generations resistance to 0·05 mg litre^{-1} total copper by the green alga *Selanastrum capricornutum* has been demonstrated over one cell generation (Kuwabara & Leland, 1986), indicating that, for this organism at least, there was some innate tolerance. Rather than introduce gradually increased concentrations over several generations they used special apparatus to increase the concentration over the course of eight hours along with precautions to ensure that the adaption was not attributable to media conditioning by algal exudates (see Section 6.3.4).

If the heavy metals exert a selection pressure then the degree of resistance is likely to be related to the severity of the pollution. In a study of field populations of the widespread filamentous green algae *Stigeoclonium tenue* and *Hormidium* spp. there was a clear relationship between aqueous concentrations of zinc and the tolerance of the population (as 'tolerance index concentration') at concentrations of about 0·2 mg litre^{-1} zinc and above (Harding & Whitton, 1976; Say *et al.*, 1977). For both species increased tolerance was accompanied by morphological changes; 'basal' growth predominating over 'upright' growth in *Stigeoclonium tenue* (Harding & Whitton, 1976) and increased 'knee-joints' in *Hormidium rivulare* (Say *et al.*, 1977). In the case of *Hormidium fluitans* collected from a copper-rich site in Papua New Guinea there was a correlation between the thickness of the gelatinous sheath and the concentration of copper in the media (Sorentino, 1985) although it is unlikely to be the sole means of resistance. In *Chlorella* and *Scenedesmus* copper concentrations which partially inhibited growth had cells up to 50% larger than controls with abnormal contents and yellow-green chloroplasts (Stokes *et al.*, 1973) and zinc-tolerant strains of '*Anacystis nidulans*' tended not to separate after cell division and instead formed filaments (Shehata & Whitton, 1982).

At the cellular level, heavy metals have been observed to affect several physiological processes; their overall toxic effect may involve combinations of these or, alternatively, some of the more specific 'effects' may be consequences of basic breakdowns in intracellular organization. Increasing concentrations of copper, zinc and cadmium extended the 'lag-phase' of growth of *Selanastrum capricornutum* before exponential growth began (Bartlett *et al.*, 1974); however once exponential growth had started the growth rate was the same at zinc concentrations up to about 0·070 mg litre^{-1}. Under these circumstances the length of the lag-phase is a more sensitive indicator of metal toxicity than either the growth rate or the stationary phase density (Kuwabara & Leland, 1986) and it may indicate a short-term adaptation of the algae to the heavy metals.

Several workers have observed a reduction of photosynthesis in the presence of heavy metals, in both algae (Fängström, 1972; Malanchuk & Gruendling, 1973; Arndt, 1974) and aquatic higher plants (Brown &

Rattigan, 1979; Rabe *et al.*, 1982) and suggested this as their principal toxic effect. The mode of action may involve replacement of the central magnesium atom of the chlorophyll molecule by the heavy metal (Arndt, 1974); however respiration has also been shown to be affected, both by direct measurement (cadmium on *Euglena*; Bonaly *et al.*, 1986) and by observations of deformed mitochondria (cadmium on *Stigeoclonium*; Silverberg, 1976). No effect on respiration was observed in *Lemna minor* exposed to elevated concentrations of copper over twenty days although photosynthesis decreased and photorespiration increased over the same period (Filbin & Hough, 1979). Damage to membranes (usually measured as leakage of potassium or other electrolytes) which has been observed (McBrien & Hassall, 1965; Singh & Yadava, 1986) may be a result of a breakdown of cellular ion pumps caused by basic metabolic failures. The same would apply to inhibition of nutrient uptake (Kashyap & Gupta, 1982; Singh & Yadava, 1983, 1984; Peterson & Healey, 1985); although this does not mean that these parameters are not effective measures of metal toxicity under many circumstances. Several of the effects observed by Filbin and Hough (1979) on copper-stressed *Lemna minor* (including decreases in photosynthesis and heterotrophic glucose uptake and increased release of organic carbon) could be interpreted in terms of damage to membrane integrity. Other effects which have been observed include inhibition of cell division in marine and freshwater phytoplankters exposed to copper (Stauber & Florence, 1987) and inhibition of the germination of seeds of rice, *Oryza sativa*, by lead (Mukherji & Maitra, 1976, 1977).

As penetration of the cell membrane is a prerequisite for heavy metals to exert a toxic action it is not surprising, perhaps, that one common mechanism of metal tolerance is to exclude the metal from the cell. This was eloquently demonstrated on strains of *Chlorella vulgaris* isolated from non-polluted ($< 0{\cdot}002$ mg litre^{-1} copper) and copper-polluted ($0{\cdot}12$ mg litre^{-1} copper) stretches of the River Hayle, Cornwall (Foster, 1977). The non-tolerant strain accumulated between five and ten times more copper than the tolerant strain (Fig. 6.2); however both strains had identical growth rates at equal amounts of cellular copper (Fig. 6.3) which, she suggested, implied that exclusion was the main mechanism of copper tolerance in this organism (Foster, 1977). Metal exclusion by tolerant strains has also been demonstrated in the blue-green alga '*Anacystis nidulans*' (Singh & Yadava, 1986). The secretion of extracellular mucilages (Sorentino, 1985) and complexing agents (see Section 6.3.4) have also been suggested as mechanisms of metal tolerance; these again act to exclude the

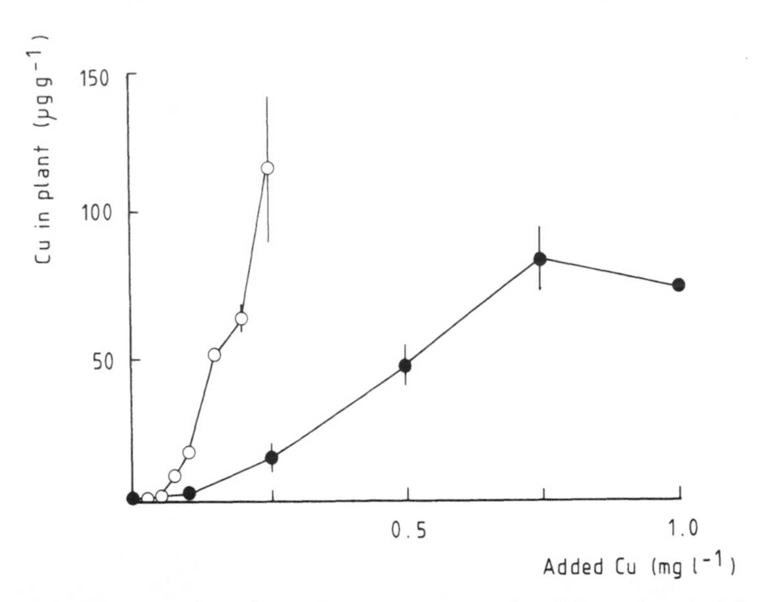

Fig. 6.2. The concentration of copper accumulated by tolerant (closed circles) and non-tolerant (open circles) strains of the green alga *Chlorella vulgaris* at different concentrations of copper in the medium. Vertical bars represent the standard error of the mean of three replicates. (Modified from Foster (1977), *Nature (London)*, **269**, 322–3 and reproduced with permission from Macmillans Journals Ltd.)

metal tolerance; these again act to exclude the metal from the cell. Metal that does penetrate the cell membrane has been observed in polyphosphate granules of blue-green algae (Jensen *et al.*, 1986; see Section 6.3.3) and in the vacuoles (Silverberg, 1975). In this study, lead at the cell membrane of *Stigeoclonium tenue* was transferred by pinocytosis to the vacuoles. There are also a few reports of metal-binding proteins produced by aquatic plants. The cytosol of *Eichhornia crassipes* subjected to 1·62 mg litre^{-1} cadmium for ten days contained two proteins with molecular weights of 2300 and 3000 associated with the accumulated cadmium (Fujita, 1985). These, like the metallothioneins, contained a high proportion of the amino acid cysteine (Fujita & Kawanishi, 1986).

6.3 ENVIRONMENTAL FACTORS AFFECTING METAL TOXICITY

6.3.1 pH

The environmental variable most often considered in addition to the

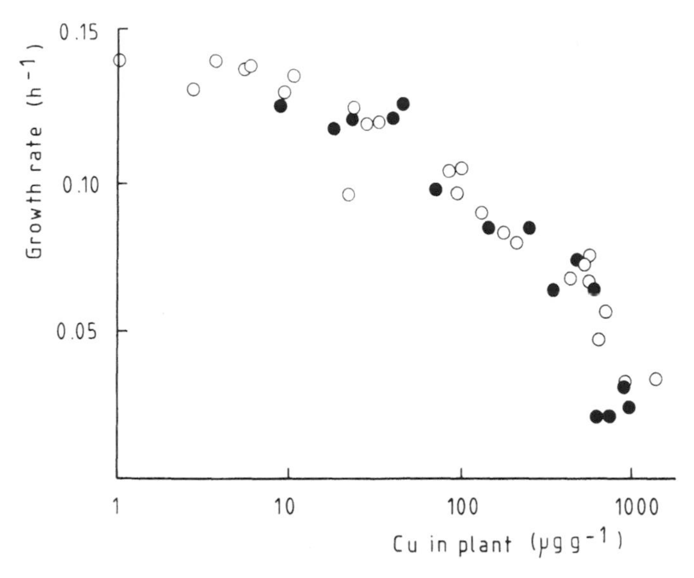

Fig. 6.3. The relationship between the growth rate (as reciprocal of the doubling time, in hours) of each culture and the amount of copper accumulated by tolerant (closed circles) and non-tolerant (open circles) of *Chlorella vulgaris* grown at various external copper concentrations. (Modified from Foster (1977), *Nature (London)*, **269**, 322–3 and reproduced with permission from Macmillans Journals Ltd.)

concentration of the metal itself is pH. Whilst many workers have measured the effects of a change in pH it is only recently that attempts have been made to separate the effects resulting from a change in metal speciation from changes in the organisms' response to the metal (Campbell & Stokes, 1985). Application of these findings may help to unravel some of the apparent confusion in the literature.

The most common observation has been for toxicity to decrease as the pH decreases (= increase in H^+ concentration); however the extent of this can vary even between different strains of the same species. Zinc-tolerant strains of *Stigeoclonium tenue* showed large differences in toxicity over the pH range 6·1–7·6 whilst non-tolerant strains showed only a slight decrease over this range (between pH 7·1 and 7·6; Harding & Whitton, 1977). In '*Anacystis nidulans*' an even more pronounced effect was observed; the tolerant strain showed the expected increase in toxicity as the pH rose whilst the non-tolerant strain showed a decrease in toxicity as pH increased (Shehata & Whitton, 1982). Amongst the higher plants, toxicity of copper to *Lemna paucicostata* was lower at pH 4·1 compared to pH 5·1 although there was no difference in the concentrations accumulated at these two pH values (Tanaka *et al.*, 1982). In contrast Rai and Kumar (1980), working on a strain of *Chlorella vulgaris*, observed the opposite effect, the greatest toxicity towards zinc at the lowest pH values, maximum growth at pH 8·0 and some inhibition again at pH 9·0 and 10·0.

The role of pH in determining chemical speciation has already been discussed (Chapter 3). Many workers consider that it is the ionic form of the metals which is the most toxic; hence any change in speciation which increases the amounts of free ions may be expected to increase the apparent toxicity of the metal. In addition, decreases in pH will decrease the proportion adsorbed onto particulates. The extent to which this will change the observed toxicity will depend upon the metal. Theoretical calculations based upon a simple inorganic system illustrated some of these differences (Table 6.1). Nickel and zinc both exist predominantly in their ionic form throughout the pH range studied (pH 4–7) whilst for copper and lead there were significant changes from the free ion towards hydroxides and carbonates as the pH rose (Campbell & Stokes, 1985). Addition of organic ligands (simulating a fulvic acid) had no effect upon the speciation of all the metals tested except for aluminium and copper, both of which were more highly-complexed at higher pH values (Campbell & Stokes, 1985). In other words pH-induced changes in toxicity, due to changes in speciation may be expected only for a few metals, including copper and, to a lesser extent, lead.

There are two components in the effect of pH on organisms' responses to metals. First, if a decrease in pH causes physiological stress then this may act synergistically with the metal to give an apparent increase in toxicity (Starodub *et al.*, 1987). On this basis one may expect a decrease in toxicity close to the pH optimum of the organism. This was shown for a population of *Hormidium rivulare* from an acid mine drainage stream which was most tolerant to zinc and copper at pH 3·5 (Hargreaves & Whitton, 1976*b*), the pH at which it also grew best (Hargreaves & Whitton, 1976*a*). The toxicity of zinc increased above and below this. Second, the mechanisms of metal accumulation are also strongly dependent upon pH and if less metal is accumulated at low pH values then this may, it follows, have less of a toxic effect. Identification of clear effects of this latter type is complicated by problems associated with concomitant changes in metal speciation. Campbell and Stokes (1985) overcame this by concentrating their attention upon those metals identified as least likely to change speciation in solution over the pH range of interest and were able to demonstrate evidence for competition between hydrogen ions and metal ions for binding sites on algal cell surfaces. Most of these studies con-

Table 6.1

Effect of acidification on the calculated speciation of dissolved trace metals in a typical surface water from the Canadian Shield (no organic ligands; redox potential, $p\varepsilon = 12$), calculated using the MINEQL-1 model (Westall *et al.* (1976))

Metal	*Total concentration* (μM)	*Percent aquo ion* $M^{z+}(H_2O)_n$				*Other dissolved species* ($\geqslant 1\%$)	*Solids*
		pH 7	*pH 6*	*pH 5*	*pH 4*		
Ag	0·01	99	99	99	99	AgCl	
Al	5·89	< 1*	< 1*	< 2*	58	AlF^{2+}, AlF_2^+, $AlOH^{2+}$	$Al(OH)_3$
Cd	0·10	99	100	100	100		
Co(II)	0·01	100	99	100	100		
Cu	0·10	37	96	100	100	$CuOH^+$, $Cu(OH)_2^0$, $CuCO_3^0$	
Hg	0·001	< 1	< 1	< 1	< 1	$HgCl^+$, $HgCl_2^0$, $HgClOH^0$, $Hg(OH)_2^0$	
Mn	0·72	< 1*	50*	100	100		MnO_2
Ni	0·10	99	100	100	100		
Pb	0·005	36	91	100	100	$PbOH^+$, $PbCO_3^0$	
Zn	0·50	99	100	100	100		

An asterisk indicates the presence of a precipitated solid (see last column); % = ([aquo ion] ÷ [M]T) × 100. Reprinted from Campbell and Stokes (1985), *Canadian Journal of Fisheries and Aquatic Science*, **42**, 2034–49 with permission from the Canadian Government.

sidered the cell walls; however in *Chlamydomonas variabilis* grown in batch culture lower pH values led to decreased adsorption of zinc onto the cell wall and a decreased flux across the cell membrane (Harrison *et al.*, 1986). This kind of relationship still exists but is less clear when the metal undergoes pH-dependent changes in speciation. Under these circumstances measurement of the total concentration of the metal is useless. In experiments using nitrogen- and phosphorus-limited cultures of *Scenedesmus quadricauda* there was a linear relationship between the concentration of the free copper ion causing 25% and 50% inhibition of ammonium and phosphate uptake and pH but no relationship when the total copper concentration was used instead (Peterson *et al.*, 1984; Peterson & Healey, 1985).

The overall effect of pH, then, will represent a 'trade-off' between the changes in speciation (making the metal more or less 'available') and changes in the organisms' response to the metal. Campbell and Stokes (1985) outlined two contrasting types of behaviour to a decrease in pH:

(i) if there is little change in speciation and metal binding at the biological surface is weak then the decrease in pH will decrease toxicity due to competition for binding sites from hydrogen ions (type I behaviour);
(ii) where there is a marked effect on speciation and strong binding of the metal at the biological surface then the dominant effect of acidification will be to increase the metal availability (type II behaviour).

Nonetheless interpretation of published results is difficult. Of the four metals of interest here there were sufficient data to examine the behaviour of all but nickel. Zinc fitted the type I classification quite well and the meagre data for lead conformed to the expected type II behaviour; however copper was expected to belong to type II (strong changes in speciation, strong binding to biological surfaces) but actually behaved as a type I metal indicating effective competition by hydrogen ions for the binding sites (Campbell & Stokes, 1985).

6.3.2 Water hardness

Water hardness (defined as the sum of the concentrations of calcium and magnesium salts) has a strong ameliorating effect on the toxicity of most metals towards plants. Populations of the green algae *Stigeoclonium tenue* and *Hormidium rivulare* collected from sites with high aqueous calcium were generally less tolerant to a particular zinc concentration than populations from sites with low calcium (Harding & Whitton, 1976; Say *et al.*, 1977). These observations were confirmed by laboratory experiments which also demonstrated that a rise in the concentration of magnesium and calcium tended to have a greater effect on tolerant compared with non-tolerant strains (Harding & Whitton, 1977; Say & Whitton, 1977). These effects were more pronounced with calcium. As the behaviour of the organisms towards the two metals was different the manner in which they alleviate toxicity may not be the same. Similar results have been found for zinc toxicity to *Chlorella vulgaris* (Rai & Kumar, 1980) and zinc and cadmium toxicity to '*Anacystis nidulans*' (Shehata & Whitton, 1982; Singh & Yadava, 1983, 1984). In contrast they had no effect on copper toxicity to '*Anacystis nidulans*' (Whitton & Shehata, 1982).

Like hydrogen ions, several workers have suggested competition for binding sites as the most likely mode of action of calcium in ameliorating metal toxicity (Stokes, 1983); certainly this is supported by observations on the effect of magnesium and calcium on metal accumulation (see Chapter 5). Differential effects of calcium on different metals may also be explained by this, the greater the affinity of a metal for binding sites the less the expected effect of calcium in moderating its effects. On this basis, calcium would be expected to have a considerably greater effect on the toxicity of zinc, a borderline metal with some Class A properties, than on a metal with strong Class B properties such as copper. To this extent the basis would be similar to that proposed by Campbell and Stokes (1985) for pH. An alternative explanation for the lack of effect of magnesium and calcium on copper toxicity in '*Anacystis nidulans*' suggested by Whitton and Shehata (1982) was that the copper is bound by complexing agents released by the algae (McNight & Morel, 1980; Clarke *et al.*, 1987). Nonetheless calcium did reduce uptake of lead and copper by *Nostoc muscorum* over a wide pH range (Schecher & Driscoll, 1985).

6.3.3 Phosphate

Although several workers have observed decreased metal toxicity as the concentration of phosphate increases the mechanism for this is still unclear. Like the hardness factors the magnitude of the effect varies between tolerant and non-tolerant strains of the same species. For both *Stigeoclonium tenue* (Harding & Whitton, 1977) and *Hormidium rivulare* (Say & Whitton, 1977) there was only a slight effect on non-tolerant populations but a much greater effect on tolerant populations.

As the major nutrient limiting growth in freshwaters, addition of phosphate often has a marked effect on algal growth quite separate from any alleviation of metal toxicity (Say & Whitton, 1977). Moreover there are differential effects depending upon the phosphate reserves of the algae. In '*Anacystis nidulans*' phosphate-starved populations were more sensitive to zinc than phosphate-rich populations, especially at low environmental phosphate concentrations (Shehata & Whitton, 1982). However the phosphate was less effective if it was added after the cells had been exposed to zinc and if it was supplied as an organic phosphate rather than as an inorganic salt (Shehata & Whitton, 1982). In the green alga *Selanastrum capricornutum* yield limitation caused by a low phosphorus concentration was intensified at elevated concentrations of zinc (Kuwabara, 1985). His data, illustrating that the zinc was interfering with phosphorus metabolism, along with experiments on *S. capricornutum* which showed the extent of lead toxicity to be partially dependent upon the phosphate-status of the culture (Monahan, 1976), indicated that it is not a simple cause-effect relationship. Other workers have shown phosphate uptake to be inhibited by copper (Peterson *et al.*, 1984; Peterson & Healey, 1985) and cadmium (Singh & Yadava, 1984). Under these circumstances the phosphorus-status of algal test populations and the phosphorus composition of the media become critical in determining the observed effects of metal toxicity; Say and Whitton (1977) recognised this after their experimental studies on *Hormidium rivulare* and speculated that some of the divergence between field observations and experimental studies may result from relatively high concentrations of phosphate in the media.

Amongst the blue-green algae polyphosphate bodies, which store phosphate accumulated during periods of plenty, may represent an additional storage site for heavy metals. Accumulation of copper, zinc and lead into the polyphosphate bodies of *Anabaena variabilis* has been observed using electron microanalysis techniques (Jensen *et al.*, 1986) and increases both in their number and size observed at elevated concentrations of several heavy metals in *Plectonema boryanum* (Rachlin *et al.*, 1983) and *Anabaena* spp. (Rachlin *et al.*, 1985). Under these circumstances, they suggested, the bodies may act as local centres for storage and detoxification of the metals (Jensen *et al.*, 1986).

6.3.4 Chelating agents

As it is the free ions which often represent the most toxic forms of metals the presence of any ligands which reduce their concentration will have a marked effect on the apparent toxicity of metals. In the laboratory many workers have used chelating agents such as EDTA (ethylenediaminetetraacetic acid) as analogues of naturally occurring substances which may include humic and fulvic acids, amino acids, peptides and other compounds. One such study, using *Scenedesmus quadricauda* and *Ankistrodesmus falcatus* as test organisms, showed growth to be inversely related to the calculated concentration of free nickel but independent of the total nickel concentration (Spencer & Nichols, 1983). Similarly, the effect of copper ions on *Spirodela polyrhiza* increased when the concentration of other transition metals exceeded that of EDTA (Schreinemakers & Dorhout, 1985). The assumption in this and other studies is that the chelating agents are acting outside the cell by affecting the chemical speciation of the metal. In other words, the toxic effect of the metal is to some extent masked by the presence of the EDTA up to a point where the concentration of the metal exceeds the binding capacity of the EDTA (Stokes, 1983). Once again, these chelating agents are going to have the most marked effects for metals such as copper which bind very strongly, and this binding will itself be partially dependent upon ambient physicochemical conditions. Part of the toxic effect of trivalent aluminium on *Scenedesmus*, for example, was attributed to its displacement of copper from chelators (Rueter *et al.*, 1987).

However, in addition to the naturally occurring chelating agents some species of algae (particularly blue-greens) are capable of releasing chelating agents of their own. Pioneering work on these by Fogg and Westlake (1955) used the blue-green alga *Anabaena cylindrica* to show that samples grown in the presence of extracellular products tended to be more tolerant to increased concentrations of copper than samples without these (Fig. 6.4). These products they showed to be polypeptides with a capacity to bind 0·325 mg copper per milligram of total peptide; however they concluded that the primary function of these agents was probably not to render toxic ions less toxic but to keep nutrients such as phosphate and iron in an 'available' form. Subsequent workers have identified some of these extracellular products (termed 'siderophores') as important agents in the uptake of iron by some blue-green algal species. For example *Anabaena flos-aquae* produced them whilst *Synechococcus leopoliensus* did not (McNight & Morel, 1980). In one study, again using *Anabaena* sp., the iron–siderophore complex was shown to be taken up and metabolized by the algae unlike the copper–siderophore complex which tended to remain in solution (Clarke *et al.*, 1987). The siderophores were produced in response to iron starvation and, again, their role in protecting against copper toxicity was secondary. In contrast Jardim and Pearson (1984) showed populations of *Plectonema boryanum* and *Anabaena cylindrica* exposed to high concentrations of copper to produce more complexing

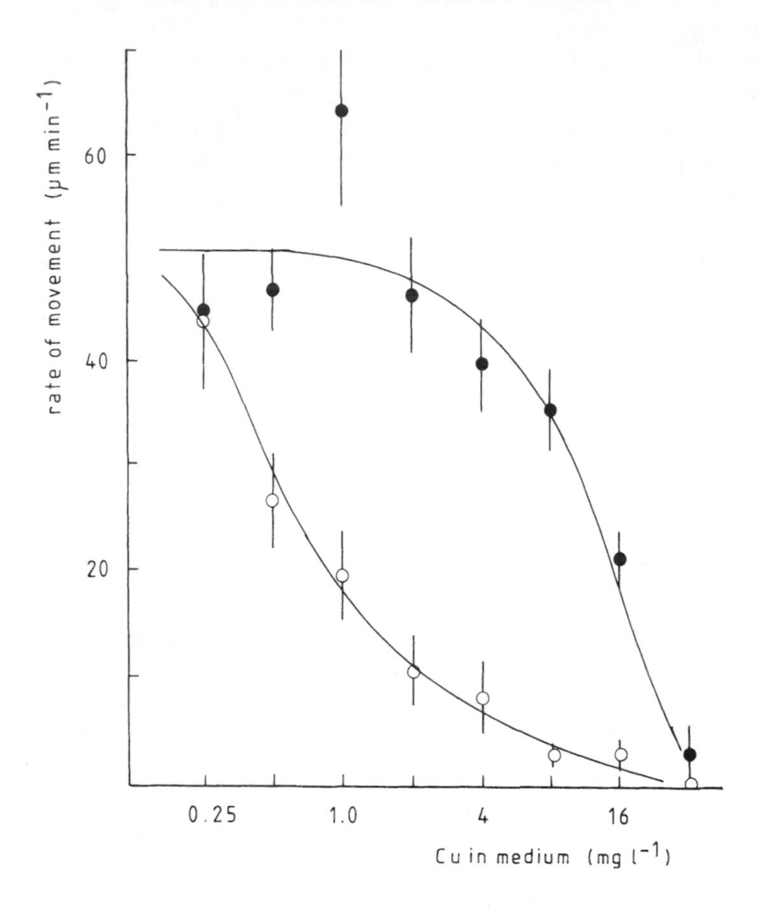

Fig. 6.4. The toxicity of copper to the blue-green alga *Anabaena cylindrica* in the absence (open circles) and presence (closed circles) of its extracellular polypeptide (7·9 mg litre^{-1} nitrogen). Toxicity was determined as the effect on the mean rate of movement of filaments, measured using a microscope. Each point is the mean of ten determinations; vertical bar is ± one standard deviation. (Redrawn from Fogg and Westlake (1955), *Verhandlungen, Internationale Vereinigung für Theoretische und Angewandte Limnologie*, **12**, 219–32 and reproduced with permission from E. Schweizerbart'sche Verlagsbuchhandlung (Nägale und Obermiller).)

agents than populations subjected to 'normal' copper concentrations. Even if their role in protecting against metal toxicity is secondary it would appear that at least some species are able to produce these compounds in response to elevated metal concentrations.

There are fewer records of these being produced by other groups of algae; recently Starodub *et al.* (1987) found a strain of *Scenedesmus quadricauda* which produced a high-carbohydrate containing compound which reduced the toxicity of copper, zinc and lead. A decrease in the pH of the medium resulted in an increase both in the toxicity of the metals and in synergism between them, presumably due to desorption of the metals from the ligands.

6.4 SYNERGISM, ADDITIVITY AND ANTAGONISM

Up to here the effects of individual metals have been considered; in practice several heavy metals are often found together in the field. Interpretation of these results is clearly going to be more complicated than the single metal scenarios that most laboratory-based studies have undertaken. Consider a mixture containing two heavy metals (mixes of more than two metals will follow the same general pattern); the combined toxic effect of these may either equal the sum of the toxicities of the two constituents (additivity), it may exceed the sum (synergism) or it may be less (antagonism). Generalizations are difficult; where four metals are being considered, there are four times four combinations. All three types of response have been observed and a few selected examples of each will be given here (Table 6.2).

A clear example of synergism was demonstrated in metal-tolerant strains of *Scenedesmus* by Stokes (1975*b*). She grew the alga in two concentrations of copper and nickel, one concentration which caused some inhibition (0·3 mg litre^{-1} copper, 1·0 mg litre^{-1} nickel) and one which caused severe inhibition (1·0 mg litre^{-1} copper, 3·0 mg litre^{-1} nickel). Even the two low concentrations combined had a greater effect on growth than the high concentrations singly. The low nickel concentration alone caused 22% inhibition of growth and the low copper concentration, 40%; however together these caused 75% inhibition of growth. Other examples of synergism include manganese and copper on *Selanastrum capricornutum* (Christensen *et al.*, 1979), zinc and cadmium on *Hormidium* spp. (Say & Whitton, 1977) and cadmium and lead in '*Anacystis nidulans*' (Whitton & Shehata, 1982). Amongst the flowering plants synergistic effects were observed between zinc and cadmium towards *Lemna minor* and *Salvinia natans* (Hutchinson & Czyrska, 1975). Zinc (0·05 mg litre^{-1}) and cadmium (0·03 mg litre^{-1}) alone produced 195 and 82 fronds of *S. natans* respectively, compared to 108 in control solutions; however when

Table 6.2
Interactions between heavy metals; effects on toxicity

Species	*Metals*	*Effect*	*Reference*
Algae			
'*Anacystis nidulans*'	Cd and Pb	Synergistic	Shehata and Whitton (1982)
Chlorella stigmatophora	Cu and Mn	Synergistic	Christensen *et al.* (1979)
Hormidium rivulare	Zn and Cd	Synergistic	Say and Whitton (1977)
Selanastrum capricornutum	Cu and Mn	Synergistic	Christensen *et al.* (1979)
Scenedesmus acutiformis var. *alternans*	Cu and Ni	Synergistic	Stokes (1975*a*)
Scenedesmus quadricauda	Cu and Zn	Additive	Petersen (1982)
'*Anacystis nidulans*'	Zn and Cd	Antagonistic	Singh and Yadava (1983)
'*Anacystis nidulans*'	Zn and Cd	Antagonistic	Whitton and Shehata (1982)
Chlorella stigmatophora	Cu and Pb	Antagonistic	Christensen *et al.* (1979)
Chlorella vulgaris	Cu and mixture of metals	Antagonistic	Wong and Beaver (1980)
Selanastrum capricornutum	Cu and Cd	Antagonistic	Bartlett *et al.* (1974)
Selanastrum capricornutum	Cu and Pb	Antagonistic	Christensen *et al.* (1979)
Higher Plants			
Lemna minor	Cu and Ni	Synergistic	Hutchinson and Czyrska (1975)
Lemna minor	Zn and Cd	Synergistic	Hutchinson and Czyrska (1975)
Salvinia natans	Zn and Cd	Synergistic	Hutchinson and Czyrska (1975)
Lemna paucicostata	Mn, Cu, Zn, Cd	Additive	Nasu *et al.* (1984*a*)
Lemna paucicostata	Cu and Cd	Antagonism	Nasu *et al.* (1986*b*)

the two metals were supplied together only 58 fronds were produced (Hutchinson & Czyrska, 1975).

Additive responses are rarer; one example was reported for copper and zinc toxicity to *Scenedesmus quadricauda* (Petersen, 1982); however this was attributed to competition between ions for binding sites on EDTA in the media rather than a direct biological response. Additive effects have also been observed between manganese, copper, zinc and cadmium toxicity towards frond production by *Lemna paucicostata* (Nasu *et al.*, 1984*a*). In contrast there are many examples of antagonistic responses of metal mixtures. For example Christensen *et al.* (1979) tested the effects of manganese (1·0 mg litre^{-1}; little toxic effect) and copper and lead (0·085 and 0·125 mg litre^{-1}, intermediate toxic effects) on *Selanastrum capricornutum* and noted antagonism between manganese and lead and between copper and lead. They observed an inhibition of cell division in the presence of lead which was partially offset by the addition of manganese. This antagonism was, in their opinion, caused by competition for active sites on enzymes. Other examples of antagonism reported include cadmium inhibition of copper toxicity in *Selanastrum capricornutum* (Bartlett *et al.*, 1974), decreased copper toxicity of *Chlorella vulgaris* in the presence of a mixture of metals (Wong & Beaver, 1980) and decreased cadmium toxicity in the presence of zinc in '*Anacystis nidulans*' (Singh & Yadava, 1983, 1984; Whitton & Shehata, 1982). Antagonism was also observed between copper and cadmium towards *Lemna paucicostata* (Nasu *et al.*, 1984*b*). In contrast to the above example (Nasu *et al.*, 1984*a*), here the effect was on flowering, a process suppressed by copper but not cadmium. When the two metals were supplied together the cadmium nullified the effect of the copper.

A cautionary note on studies of synergism and antagonism was sounded by Wong and Beaver (1981). They compared the 'conventional' approach to these studies (i.e. supplying a mixture of two metals) with a more sophisticated approach in which *Selanastrum quadricauda* was pretreated with one metal before being exposed to the second. By the conventional method lead and zinc acted synergistically; however when cells pretreated with zinc were harvested and resuspended in a medium containing lead the effect was antagonistic. In other words the result could vary depending upon the experimental protocols. Their conclusions are an appropriate point at which to end this chapter as they suggested that the conventional methods failed to separate reactions outside the cell from those within the cell, a theme which I have tried to stress throughout this chapter.

Chapter 7

Toxicity and Tolerance to Heavy Metals. II. Animals

7.1 FIELD STUDIES

Some of the earliest field studies of the effects of heavy metal mining on the freshwater fauna were performed by Kathleen Carpenter (1924, 1926) in the rivers of mid-Wales. These are interesting in their own right but are made more so by studies documenting subsequent changes (e.g. Jones, 1940, 1958) which make this area one of the most important case studies of freshwater heavy metal pollution in the literature. When Carpenter made her first studies some of the lead mines were still operational and others had only recently closed and the state of the streams must have been little different to when the mines were operational; thus there are records of the state of these rivers covering some sixty years.

One of these streams, Marchant Brook (a tributary of the Teifi) had apparently recovered from lead pollution from a mine which closed in 1918 and, in the spring of 1924, its fauna was little different from other moorland streams in the area. However, later in 1924 mining activities started again and began to discharge partially treated minewaters into Marchant Brook. When Carpenter (1926) resurveyed this stream in the summer of 1925 there were clear effects upon the biota. Gone, for example, was the freshwater limpet *Ancylus fluviatilis* and all of the species of Trichoptera whilst the larvae of Plecoptera and Ephemeridae appeared unaffected. Two and a half kilometres downstream of the mine she observed the full fauna, including fish, restored. She attributed these effects to lead brought into solution by the action of the water on the solid mine waste although it is now considered that many of the observed effects were probably caused by the more soluble zinc.

The effects of the mines were also obvious in the nearby River Rheidol.

Some of the lead mines here had reopened during the First World War and these had obvious effects on the benthos, reducing it to just 14 species of invertebrate (Carpenter, 1924); however once the last mine had closed in 1921 improvements in the water quality occurred rapidly. Before the mines closed there was on average between 0·2 and 0·5 mg litre^{-1} lead in the water; this subsequently dropped to about 0·02 mg litre^{-1} lead except during floods when it rose again. In 1922 she recorded 29 species including a platyhelminth and eight species of Trichoptera. Neither of these groups was present in the earlier survey (Carpenter, 1924). A decade later the total number of species had increased to 100 with oligochaetes, Hirudinae and a lepidopteran amongst the new taxa recorded. Also, some fish species had been observed (Laurie & Jones, 1938). By the end of the next decade Jones (1949) was able to record 190 species of which 130 were found in the main stream. By this time the benthos may have been approaching an equilibrium; Jones and Howells (1975) mention an unpublished survey which showed the state of the Rheidol to be very similar then to what Jones (1949) had observed and this is confirmed when recruitment of species to the benthos is plotted over time (Fig. 7.1). The outlying point (bracketed)

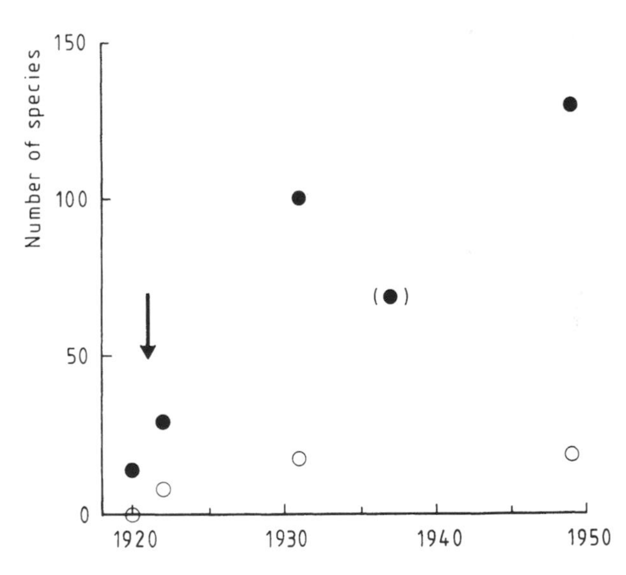

Fig. 7.1. The recovery of the invertebrate fauna in the River Rheidol, mid-Wales following the cessation of lead mining (arrowed). Closed circles = all invertebrates; open circles = Trichoptera species only; bracketed circle = sample collected after period of heavy rain. Original figure based on data in Carpenter (1924), Laurie and Jones (1938) and Jones (1949).

represents a survey conducted in 1936–37 following a wet summer and autumn which probably reduced the fauna. Sixty-eight species, including all of the common forms from their earlier survey, were collected (Laurie & Jones, 1938). Figure 7.1 also shows the increase in number of trichopteran species collected over this period; none were collected by Carpenter (1924) in her earliest survey, but there was a rapid increase in the following years.

Zinc was identified as the main toxic agent in a later study of the river Ystwyth in the same area (Jones, 1940). It was present at concentrations up to 1·2 mg $litre^{-1}$ whilst lead was found only up to 0·05 mg $litre^{-1}$. Here the fauna was restricted to platyhelminthes (flatworms), Hydracarina (water mites) and Insecta whilst there were no molluscs, malacostracean Crustacea (e.g. *Gammarus* and *Asellus*) or fish at all. However some species were abundant, even close to the mine; these included some plecopterans (especially *Leuctra* and *Nemoura*), ephemeropterans (*Rhithrogena* and *Baetis*) and the dipteran *Simulium*. Numbers of *Baetis* increased downstream. Jones (1940) attributed some of these effects to changes in the composition of the substrate. The most abundant species were those characteristic of swift water which lived on and between stones (the stone-clinging genera *Rhithrogena* is an example) whilst mud dwellers such as *Caenis* and *Ephemerella* (Ephemeroptera) along with the oligochaetes were absent. In the late 1950s the concentration of lead in the water had declined to 'negligible' whilst the zinc concentration was in the range 0·2 to 0·7 mg $litre^{-1}$. However the fauna was still broadly similar to this although *Chloroperla* was now the dominant plecopteran and brown trout had returned (Jones, 1958). The reduction of the benthos at elevated concentrations of metals appears to be a fairly general phenomenon; in a study of two neighbouring rivers in northern England there was a negative correlation between the number of species recovered and the filtrable zinc concentration in the water (Abel & Green, 1981).

Another well-studied example is the effect of zinc from a mine at Captains Flat on the fauna of the Molongo River. The Molongo River, a tributary of the Murray, rises in the Eastern Highlands Range in New South Wales, Australia and flows, via Captains Flat about 50 km to Canberra where it flows through an artificial lake, Lake Burley-Griffen. It was concern for the sport-fishing potential of this lake which provided the original impetus for the study (Weatherley *et al.*, 1967).

Mining at Captains Flat started in 1882 and continued, with interruptions, until 1962. However it was only after 1938 when the mine was reopened by a different company that the serious pollution began. Before

1938 the river still supported good trout fishing. A survey of the invertebrate fauna in 1963, shortly after the mine closed, detected an extremely impoverished fauna ($<$ five species) immediately below the mine (34·1 mg litre^{-1} zinc in 1961) and a gradual recovery downstream (Weatherley *et al.*, 1967). Taxa represented included aquatic Hemiptera, Coleoptera, Trichoptera and Chironomidae. In contrast, just above the mine thirty species were collected, representing about a dozen taxa. They attempted to separate out two groups of effects: taxa such as the Crustacea, Mollusca and Oligochaeta were directly affected by the metals whilst groups such as the Ephemeroptera, Trichoptera and Megaloptera were affected by the increased 'scour' caused by unstable physical conditions. A reduction in the flora contributes to the instability of the stream bed as well as affecting the food supply of some invertebrates (Weatherley *et al.*, 1967).

About twenty years later, despite remedial works to stabilize the spoil heaps (Craze, 1977) the concentrations of zinc in the river immediately below the mine were still very high (up to 25·3 mg litre^{-1} 'total' zinc) and there was little improvement in the invertebrate fauna (Norris, 1986). The general pattern was the same although the number, of both taxa and individuls, was much higher 58 species at control site compared with 30 in Weatherley *et al.*, 1967), attributed to the low flows and smaller mesh-opening of the sample net in the more recent study (Norris, 1986).

Another Australian river, the South Esk River in Tasmania, has also been studied twice, once in the early 1970s and again in the mid-seventies. Although tin and wolfram (tungsten) are mined in the area (both very insoluble) the main heavy metal pollutants were copper, zinc, cadmium and lead along with manganese, iron and sulphuric acid. In the original study (Tyler & Buckney, 1973) Storys Creek, the tributary which carried the minewaters, contained 0·15 mg litre^{-1} copper, 9·2 mg litre^{-1} zinc and 0·13 mg litre^{-1} cadmium in the filtrable fraction at a sampling site just above the confluence with the South Esk River with correspondingly lower concentrations in the main river itself. As in the other studies, a general depauperation of the fauna was observed, along with the complete absence of Crustacea, Annelida and Mollusca (Thorp & Lake, 1973). In contrast the Hemiptera and Arachnida did not appear to be affected. Other species, such as some trichopterans, appear to thrive, perhaps due to the absence of predators such as trout. At all of the sites there was a reduction both in species diversity and abundance during the winter; however this was particularly marked at the polluted stations, perhaps due to the increased flooding and, with it, increased scour (Thorp & Lake, 1973).

During the second survey concentrations in Storys Creek appear to have declined slightly; mean concentrations in the filtrable fraction measured over two years were 0·076 mg litre^{-1} copper, 3·67 mg litre^{-1} zinc, 0·113 mg litre^{-1} cadmium and 0·020 mg litre^{-1} lead (Norris *et al.*, 1981). The accompanying survey of invertebrates (Norris *et al.*, 1982) was considerably more detailed than the previous one (Thorp & Lake, 1973); however the general findings were very similar. Using a variety of multivariate statistical techniques three groups of organism were identified:

(i) taxa that were abundant at both contaminated and uncontaminated sites (two species; *Oecetis*, a genera of leptocerid Trichoptera and the Plecopteran *Baetis baddamsae*). *B. baddamsae* was distinct in being the only Plecopteran which was tolerant of the metal contamination;
(ii) taxa that were most abundant upstream of the contamination (two molluscs, four leptophlebiid plecopterans and five trichopterans);
(iii) taxa most abundant below the source of contamination (six dipterans, four trichopterans, one mollusc, one amphipod and one hemipteran).

In addition to the detailed analysis of their results, Norris *et al.* (1982) also calculated values of the Margalef and Shannon diversity indices for each site. Both demonstrated the gross effects of the pollution although resolution of differences between contaminated sites was more difficult. Nonetheless there were some correlations between the indices and metal concentrations in the water and sediment.

There were also significant differences in the drift fauna between the uncontaminated and contaminated sites. At the uncontaminated sites the dominants in the drift fauna comprised Ephemeroptera, Plecoptera and Trichoptera whilst at contaminated sites there was a reduction in the Ephemeroptera and increases in the Coleoptera and Hemiptera (Norris *et al.*, 1982). Some of these changes may be attributable to behavioural responses; in the King River, another Tasmanian river polluted by copper, zinc and lead, the clear nocturnal peak in drift fauna was greatly elevated immediately below the source of contamination, especially by ephemeropterans (Swain & White, 1985) which they interpreted as organisms attempting to leave the contaminated area and, they speculated, re-enter the benthos further downstream. At two lower sites the nocturnal peaks were reduced, presumably as a consequence of the reduced benthic fauna.

Much of the work on the faunas of polluted streams has concentrated on organic pollution. Some of the indices derived to monitor this have also been applied to streams receiving heavy metals despite some reservations as to their suitability for this (Hellawell, 1978). In Willow Brook in Northamptonshire, a stream with very hard water (500 mg litre^{-1} $CaCo_3$ total hardness) both the Trent and Chandler biotic indices were applied to invertebrate communities below a steel works (Solbé, 1977). At the most polluted site ($>$ 25 mg litre^{-1} zinc) there were abundant tubificid worms and chironomid larvae and 3·4 km downstream of this the water-louse *Asellus* reappeared. Further downstream still less tolerant invertebrates increased (but not Ephemeroptera, Plecoptera or *Gammarus*) and the number of fish species and their total biomass increased. However in another study, this time in the River Nent system in northern England, the Chandler index was a poor monitor of dissolved zinc concentrations (Armitage, 1979). In both of these studies the substrate was an important factor in the response of the fauna to the metals; in Willow Brook the riffle fauna was more sensitive than the silt fauna (Solbé, 1977) whilst in the River Nent, Armitage (1979) suggested that the massive growths of green alga *Stigeoclonium tenue* prevented colonization by stone-clingers such as the ephemeroptera *Rhithrogena semicolorata* or by juvenile plecopterans whilst favouring such dipterans as the Orthocladiinae.

In a further study on the River Nent system, Armitage and Blackburn (1985) were able to distinguish between sites with different degrees of zinc pollution on the basis of the Chironomidae fauna alone. There was a negative relationship between the number of chironomid taxa and the zinc concentration in the water but not between their total abundance and zinc concentration. When a clustering technique was applied to these data similar results were obtained to those for the whole invertebrate fauna (Armitage, 1979). A similar study was performed in the River Mogami in northern Japan, a river polluted by copper, zinc, cadmium and lead (Yasuno *et al.*, 1985); however they were able to identify three species of Orthocladiinae which were very abundant even at the most polluted reaches, probably because of a lack of either competitors or predators. Orthocladiinae chironomids also dominated the metal-polluted Slate River in the USA (LaPoint *et al.*, 1984). These are valuable results for two reasons; first, there was the possibility that preliminary monitoring of sites where sampling the benthos was difficult could be carried out using just the drifting pupal exuviae (Armitage & Blackburn, 1985) and second, because the chironomids are so widespread and are often abundant in organically polluted rivers where few other invertebrate taxa survive.

Several of these studies identified the Hirudinae (leeches) as a group which were fairly sensitive to heavy metals (e.g. Laurie & Jones, 1938;

Norris *et al.*, 1982). The effect of zinc on one species of leech, *Erpobdella octoculata*, studied by Willis (1985) was a general reduction in its ecological fitness. In comparison to populations above, populations in the Afon Crafnant in North Wales below the input showed a delay in cocoon deposition and, hence, hatching and more misshapen and empty cocoons as well as a smaller proportion of young to adult leeches. There was also some evidence of lower densities which, he suggested, was related to the reduced availability of food at this lower site.

There have been fewer field surveys of fish populations in metal-polluted rivers (in contrast to laboratory studies); however differential tolerance to the metals, coupled with changes in the food supply and the molar action of the rivers has been demonstrated to affect fish distribution. In Willow Brook, the most polluted zone was found to be fishless. The first species of fish to return as the concentrations decreased downstream was the stickleback (*Gasterosteus aculeatus*) followed by the roach (*Rutilus rutilus*), tench (*Tinca tinca*) and eel (*Anguilla anguilla*), then the bullhead (*Cottus gobio*), gudgeon (*Gobio gobio*) and stone loach (*Noemacheilus barbatulus*) and finally the rainbow trout (*Salmo gairdneri*), chub (*Leuciscus cephalus*), minnow (*Phoxinus phoxinus*) and dace (*Leuciscus leuciscus*) (Solbé, 1977). In the River Rheidol, mentioned above, the first records of fish were of *Gasterosteus aculeatus*, made by Carpenter in 1925. When it was resurveyed in 1931–32 *G. aculeatus* and *Anguilla anguilla* were collected and *Salmo trutta* was observed (Laurie & Jones, 1938). Sea trout (migratory *S. trutta*) and *Salmo salar* followed, with the first definite record of *S. salar* in 1952, since when the river has been regularly restocked (Jones & Howells, 1975).

To summarize all of these field surveys involving a variety of metals in different combinations is clearly difficult; however two general points emerge:

(i) as for plants (Whitton & Diaz, 1980) there is a general reduction in the numbers of species recorded as the aqueous concentration of metals increases;
(ii) certain taxa (in particular, Oligochaeta, Mollusca, Crustacea, Trichoptera, *Salmo*) appear to be more sensitive to heavy metals than others.

7.2 EXPERIMENTAL STUDIES

There have been a great number of experimental studies on heavy metal pollution. Like so many areas of science this productivity on the part of scientists is not necessarily a key to a greater understanding of the subject and, with information on the toxicity of substances to aquatic organisms an essential prerequisite to legislation, there is a great need for standardization. In a recent review (Mance, 1987) the need to standardize a wide range of physical, chemical and biological parameters in order to make the tests reproducible was emphasized. However any appraisal of the literature shows what a large variety of techniques are employed, making comparisons between some of these reports difficult. Moreover many of these studies emphasize how the traditional criteria used to assess toxicity may ignore many chronic symptoms which reduce the organisms' ecological fitness.

7.2.1 Invertebrates

Assessed solely on the basis of acute toxicity (LC_{50}) copper is probably the most toxic of the four metals under particular consideration here (Fig. 7.2), with toxic effects apparent in some organisms at aqueous concentrations below 0·01 mg litre^{-1}. For nickel, zinc and lead there are only a few records of acute toxicity at concentrations less than 0·1 mg litre^{-1}. These figures, however, only give a very approximate picture and ignore differences between taxa and in the experimental protocols used by different workers. When different taxonomic groups are compared for their relative tolerances to copper and zinc, the two metals for which the most information is available (Figs. 7.3 and 7.4) it is clear that the Insecta are generally more tolerant to both metals than either the Mollusca or the Crustacea with the Annelida showing a very wide range of responses to zinc but remaining less tolerant to copper than either the Mollusca or the Crustacea. These results substantiate observations made in the field (Section 7.1); however the concentrations which cause 50% acute toxicity under these controlled conditions are often rather higher than the concentrations where effects are observed in the field confirming the importance of non-acute toxic effects stressed above.

An elegant demonstration of this used mixed cultures of *Daphnia pulex*, a cladoceran which is sensitive to copper, and *D. magna*, which is able to develop copper tolerance (LeBlanc, 1985). *D. pulex* is usually the superior competitor for food (in this case, *Selanastrum*) and this was shown both in the absence of copper from the medium and after a temporary exposure to 0·01 mg litre^{-1} copper for 14 days. However after temporary exposure to 0·03 mg litre^{-1} and sequential exposure to 0·01 followed by 0·03 mg litre^{-1} copper *D. magna* dominated and there was a reduction in the initial population size of *D. pulex*. Once it had recovered from the temporary

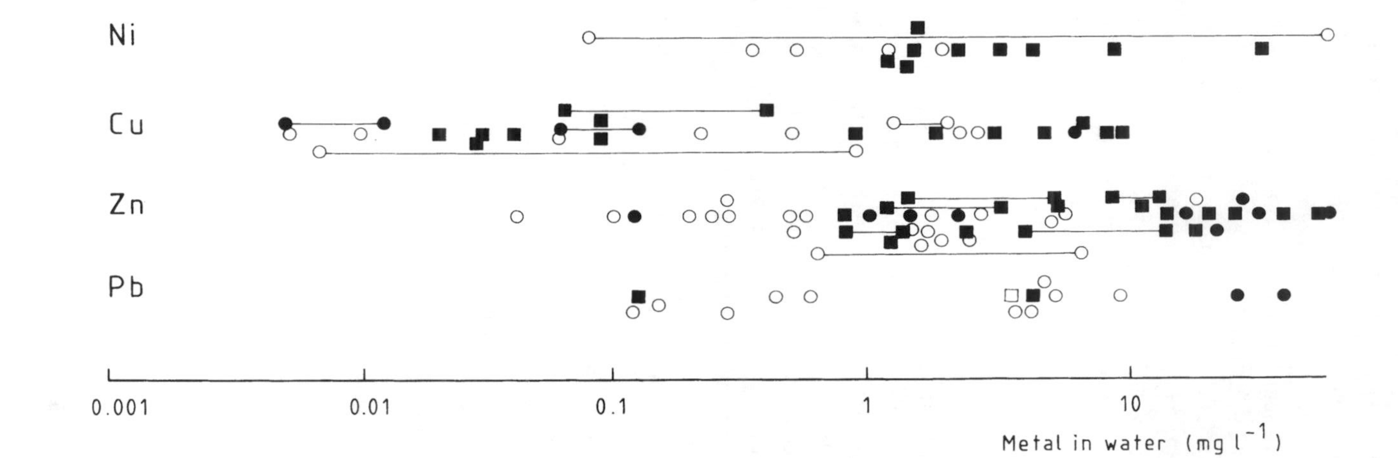

Fig. 7.2. Effect of nickel, copper, zinc and lead on aquatic invertebrates expressed as LC_{50}. No attempt is made to distinguish between different taxa; however distinctions have been drawn between 24-h LC_{50} (closed circles), 48-h LC_{50} (open circles), 96-h LC_{50} (closed squares) and LC_{50} values based on incubation times in excess of ten days (open squares). Solid lines indicate range of effects obtained from different treatments. Original figure compiled from values in the literature (see Appendix 2).

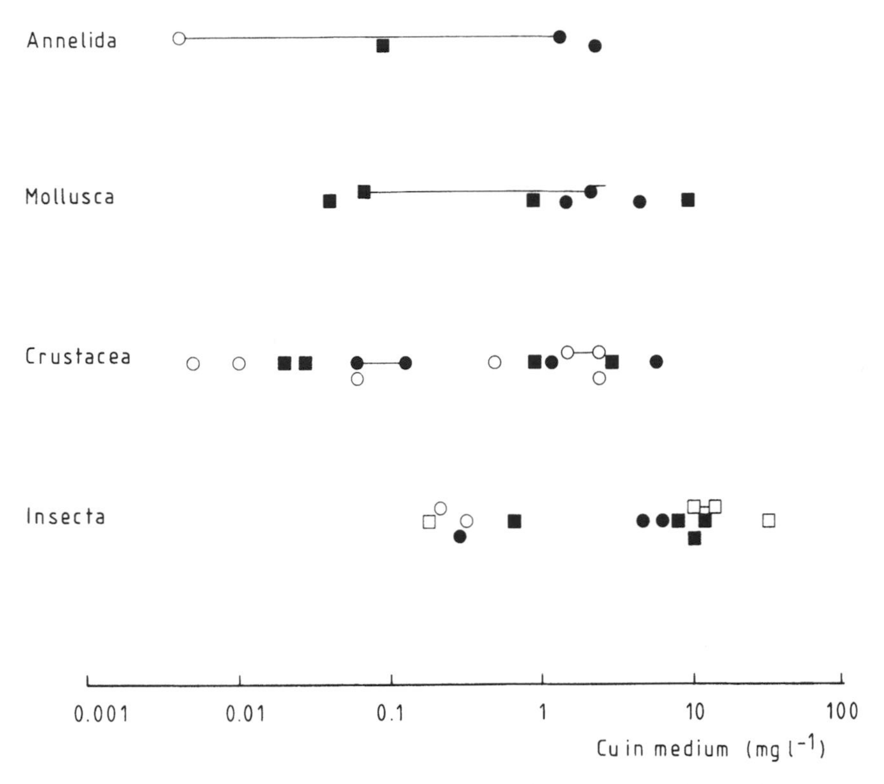

Fig. 7.3. Comparative effect of copper on four groups of aquatic invertebrates, expressed as LC_{50}. Details as for Fig. 7.2. Original figure compiled from values in the literature (see Appendix 2).

exposure to copper, however, the *D. pulex* population was able to increase and eventually to regain its dominance (LeBlanc, 1985). Changes in the fitness of *D. magna* in response to chronic concentrations of nickel (0·05 mg litre^{-1}) and copper (0·02 mg litre^{-1}) included a decline in fecundity, prolonged maturation and decreases both in the number and size of broods for nickel, and low fecundity, impaired reproductive ability, and shorter life spans for copper (Lazareva, 1986).

Part of the apparent reduction in ecological fitness results from a greater sensitivity to metals in the early life stages of many organisms. The reduction in fitness of a population of *Erpobdella octoculata* (Willis, 1985) as measured in the field has already been mentioned; similar effects have also been demonstrated in the laboratory. For example the effect of 0·005 mg litre^{-1} copper to the isopods *Asellus aquaticus* and *Proasellus coxalis* was greatest on the embryonic and larval stages (Nicola Giudici *et*

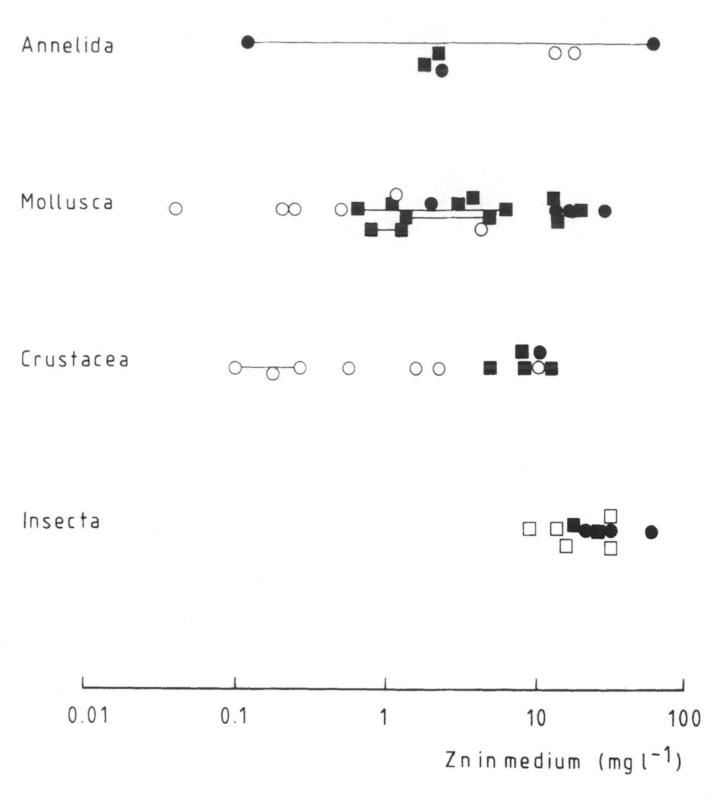

Fig. 7.4. Comparative effect of zinc on four groups of aquatic invertebrates, expressed as LC_{50}. Details as for Fig. 7.2. Original figure compiled from values in the literature (see Appendix 2).

al., 1987). Sensitivity to heavy metals appears to decrease as the organism approaches maturity; juveniles of the gastropod *Potamopyrgus jenkinsi* had a 96-h LC_{50} of 0·054 mg litre^{-1} copper whilst in adults it was 0·079 mg litre^{-1} (Watton & Hawkes, 1984). These effects have been particularly well documented for insect (particularly Chironomidae) larvae, where the juvenile stages are clearly separated into 'instars' by ecdysis ('moulting' of the exoskeleton). The 96-h LC_{50} for *Chironomus tentans* larvae to copper was between 12 and 27 times higher for first instar compared with the fourth instar (Gauss *et al.*, 1985); similar effects have also been shown for *C. riparius* to nickel (Powlesland & George, 1986) and cadmium (Williams *et al.*, 1986). The emergence of adult chironomids was both delayed and reduced from pupae exposed to contaminated sediment from Palestine Lake, Indiana (1680 μg g^{-1} chromium; 15100 μg g^{-1} zinc; 1070 μg g^{-1}cadmium) when compared with control sediments similar in texture and organic matter content (Wentsel *et al.*, 1978*a*); however other workers have detected little effect of metals to chironomids at this stage (Gauss *et al.*, 1985; Powlesland & George, 1986) although very different methods were used in these three experiments.

In the presence of contaminated sediments from Palestine Lake along with control sediments non-tolerant chironomid larvae tended to select control in preference to the contaminated sediments (Wentsel *et al.*, 1978*b*). When the tubificid oligochaetes *Tubifex tubifex* and *Limnodrilus hoffmeisteri* were exposed to sediments contaminated with copper and zinc they both attempted to move away to surrounding, uncontaminated sediments (McMurtry, 1984). Similar behavioural responses have also been shown in experiments using *Gammarus pulex* in choice chambers where the proportion of time spent in the zinc solution decreased as its concentration increased (Abel & Green, 1981). At the same time there was an increase in the proportion of time that *G. pulex* spent actively swimming.

The effect of copper and zinc on the freshwater shrimp *Macrobrachium carcinus* was to reduce the rates of respiration and ammonia excretion (Correa, 1987). This in turn had the effect of reducing the ratio of oxygen to nitrogen in the organism and increased its dependence upon carbohydrate and lipid reserves. A reduction in respiration was also observed in the crayfish *Orconectes rusticus* exposed to copper although here there were at least two components to the effect: at 0·63 mg litre^{-1} copper there was a reduction in the organism's ability to use added succinate as an energy source; however there was no sign of a breakdown in its endogenous respiration below 63 mg litre^{-1} (Hubschman, 1967*b*). This latter figure seems very high in comparison with other figures in the literature. Hubschman (1967*b*) also noted breakdown in organization of cells in the antennal gland, which is involved in osmotic and ionic regulation. During the course of 30 days exposure to 0·5 mg litre^{-1} copper, first vacuolation of the cells in the labyrinth and eventually complete destruction of the gland was observed; these effects were not apparent after short periods of exposure to copper. In another species of *Macrobachium*, *M. lamarrei*, exposure to toxic solutions of copper and zinc caused profuse secretions of mucilage by the organism (Murti & Shukla, 1984) whilst the freshwater snail *Lymnaea luteola*, exposed to its LC_{50} concentration of copper (2 mg litre^{-1}) increased its production of fatty acids (Mal Reddy & Venkateswara Rao, 1987) which, the authors speculated, were responsible for complexing the copper prior to storage or detoxification.

As for plants, so prolonged exposure of invertebrates to heavy metals can lead to the development of tolerance. In Palestine Lake, Indiana Wentsel *et al.* (1978*b*) noted significantly greater survival (75% compared to 47·5% in controls) in *Chironomus tentans* collected from contaminated sediments, coupled with less avoidance activity which, they suggested, implied that the tolerance was due to innate resistance to the metals. Populations of the isopod *Asellus meridianus*, one collected from the copper-polluted River Hayle in Cornwall and the other from the nearby lead-polluted River Gannel, were both tolerant to lead but only the river Hayle population was tolerant to copper (Brown, 1977*c*). This was associated with differential accumulation of lead and copper by the two populations: as the River Hayle population accumulated lead so the concentration of copper in the organism decreased whilst in the River Gannel population copper accumulation was lower and unaffected by changes in the ambient lead concentration. Conversely, as copper was accumulated by the River Hayle population so the amount of lead accumulated decreased. However, for both populations the effect was not as marked as for copper which suggested that the lead may have been more preferentially bound than copper (Brown, 1978). Subsequent histological work located both copper and lead in 'cuprosomes' in the hepatopancreas which acted to store and detoxify the metals. Lead was bound to these at the expense of copper which explained the reciprocal changes in whole body concentrations (Brown, 1978).

There are also many records of heavy metal toxicity in invertebrates being affected by environmental factors such as pH and water hardness. Rather than repeat much of what was written in Chapter 6 only a brief summary will be given here; in particular, for information on the effects of pH on toxicity readers are again referred to the excellent review by Campbell and Stokes (1985). Strong reductions in toxicity with increasing water hardness have been oberved for different instars of the larvae of *Chironomus tentans* and copper (Gauss *et al.*, 1985), and for *Daphnia magna* and *D. pulex* to copper, zinc and cadmium (Winner & Gauss, 1986). Both Winner and Gauss (1986) for *Daphnia* and Wright (1980) for *Gammarus pulex* related the effects of calcium on accumulation and toxicity of heavy metals. For cadmium accumulation by *Gammarus*, which was mostly internal, there was a negative relationship between calcium and cadmium which suggested that the two ions were competing for the same uptake pathway (Wright, 1980). Whilst decreased accumulation of the heavy metals in hard water was also observed in the study on *Daphnia* the relationships were far from straightforward which, they suggested, was probably as a result of the complex storage, transformation and excretion mechanisms in multicellular animals (Winner & Gauss, 1986).

Humic acid, a naturally occurring chelating agent, also had an ameliorating effect on the toxicity of copper, zinc and cadmium to daphnids (Winner, 1984; Winner & Gauss, 1986). In the absence of humic acid the 72-h LC_{50} to *Daphnia* spp. was 0·028 mg $litre^{-1}$ copper whilst in the presence of 1·5 mg $litre^{-1}$ humic acid this was increased to 0·053 mg $litre^{-1}$ (Winner, 1984). Other workers have examined the effects of suspended solids on the toxicity of zinc to *Daphnia magna*. They used three kinds of suspended solid: two sediments from local lakes and montmorillonite clay. At 1·0 mg $litre^{-1}$ zinc there was a reduction in toxicity only at 100 mg $litre^{-1}$ suspended solids and greater whilst at 20 mg $litre^{-1}$ zinc only those solids which increased alkalinity, hardness and total dissolved carbon in the water had any significant effect (Hall *et al.*, 1986). They attributed both of these effects to changes in the chemical partitioning of the zinc.

In the freshwater snail, *Lymnaea luteola*, there were also seasonal fluctuations in sensitivity to zinc which was attributed to changes in water temperature. Based on the criterion of failure to respond to a stimuli applied to the foot the 96-h LC_{50} at 17·5°C was 11 mg $litre^{-1}$, dropping to 1·68 mg $litre^{-1}$ at 32°C (Khangarot & Ray, 1987*b*).

Finally, there are also examples of competitive interactions between heavy metals (Table 7.1). For a freshwater shrip *Paratya tasmaniensis* zinc and cadmium appeared to behave antagonistically at low concentrations but at higher concentrations their effect was additive (Thorp & Lake, 1973). Copper and zinc also behaved additively in their effects on avoidance behaviour by *Tubifex tubifex* and *Limnodrilus hoffmeisteri* (McMurtry, 1984).

7.2.2 Fish

As for the invertebrates it is instructive to compare the relative toxicities of each of the four metals at the outset (Fig. 7.5). Once again copper proves to be the most toxic of the four metals, followed by zinc and lead whilst there are few records of nickel toxicity at concentrations below 10 mg $litre^{-1}$. Compared with other fish the Salmonidae appear to be particularly sensitive to copper (Fig. 7.6) and, to a lesser extent, zinc (Fig. 7.7). Again, these results are borne out by observations in the field (Weatherley *et al.*, 1967; Solbé, 1977).

One of the simplest methods of assessing the toxic potential of a water body is to expose them to the water directly using cages. This was one of the earliest toxicity testing methods used; Carpenter (1924) left two brown trout *Salmo trutta* in a cage in the River Rheidol just above Capel Bangor

Table 7.1
Some examples of interactions between heavy metals; effects on toxicity to invertebrates and fish

Species	*Metals*	*Effect*	*Reference*
Invertebrates			
Copepods	Cu and As	Synergistic	Borgman (1980)
Paratya tasmaniensis	Zn and Cd (low concentrations)	Antagonistic	Thorp and Lake (1973)
	Zn and Cd (high concentrations)	Additive	Thorp and Lake (1973)
Tubifex tubifex	Cu and Zn	Additive	McMurtry (1984)
Limnodrilus hoffmeisteri	Cu and Zn	Additive	McMurtry (1984)
Fish			
Salmo gairdneri	Cu and Zn	Synergistic	Lloyd (1960, 1961)
S. gairdneri	Cu and Zn	Synergistic	Sellers *et al.* (1975)
Lepomis macrochirus	Cu and Zn	Additive	Thompson *et al.* (1980)
Salmo gairdneri	Cu and Zn	Additive	Lloyd (1961)
S. gairdneri	Ni, Cu and Zn	Additive	Brown and Dalton (1970)
S. gairdneri	Cu and phenol	Additive	Brown and Dalton (1970)
S. gairdneri	Cu, Zn and phenol	Additive	Brown and Dalton (1970)
Catostomus commersoni	Cu and Al	Antagonistic	Markarian *et al.* (1980)
C. commersoni	Ni and Al	Antagonistic	Markarian *et al.* (1980)

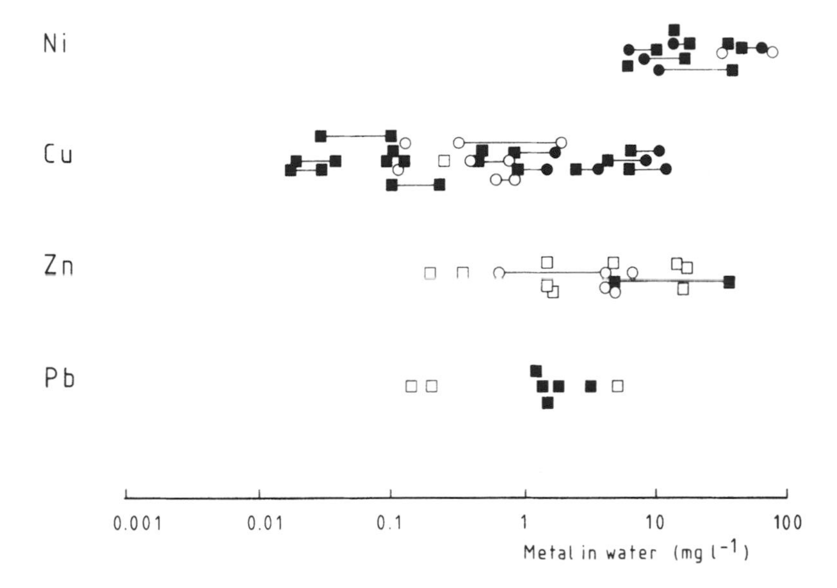

Fig. 7.5. Effect of nickel, copper, zinc and lead on fish expressed as LC_{50}. No attempt is made to distinguish between different taxa; however distinctions have been drawn between 24-h LC_{50} (closed circles), 48-h LC_{50} (open circles), 96-h LC_{50} (closed squares) and LC_{50} values based on incubation times in excess of five days (open squares). Solid lines indicate range of effects obtained from different treatments. Original figure compiled from values in the literature (see Appendix 2).

in the spring of 1924. For 18 days of low or medium flows no lead was detectable in the river and both fish survived; however following two days of heavy rain the concentration of lead (and, presumably, zinc) rose and both fish died (with a film of white mucous covering the gills) confirming the intermittent nature of the pollution in this river. More recently, similar tests have been performed in the Molongo River in New South Wales. In these *S. trutta* proved to be considerably more tolerant to the elevated zinc concentrations in the river than rainbow trout (*S. gairdneri*; Weatherley *et al.*, 1967). These experiments were repeated after the remedial work on the tailings dumps performed after the first study (see Section 7.1); however there appeared to be little positive effect on the water quality as a result. High proportions of *S. gairdneri* still died at the polluted site compared with the control sites (Graham *et al.*, 1986). After 12 days in the cages, however, even the fish at the control site appeared to have suffered from their time in the cages, particularly with anterior skin abrasions which in turn may reduce their resistance to stress. Based on

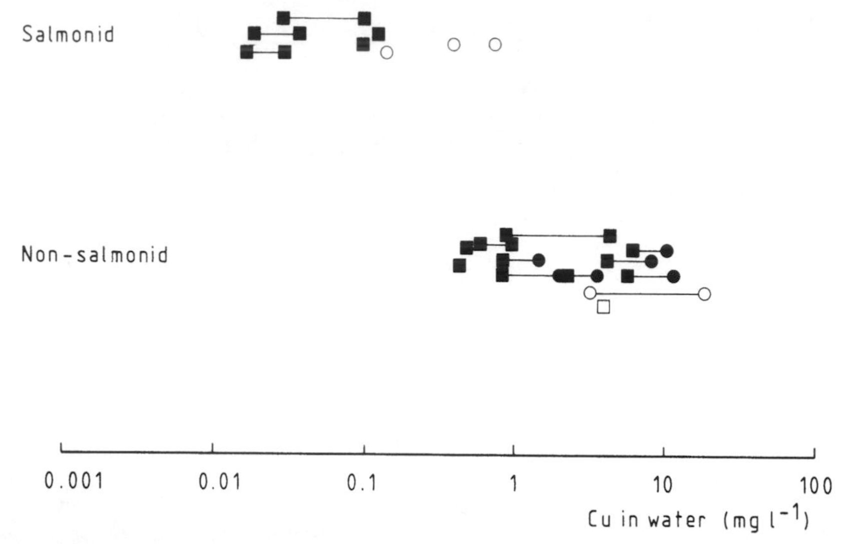

Fig. 7.6. Comparative effects of copper on Salmonidae and other fish, expressed as LC_{50}. Details as for Fig. 7.5. Original figure compiled from values in the literature (see Appendix 2).

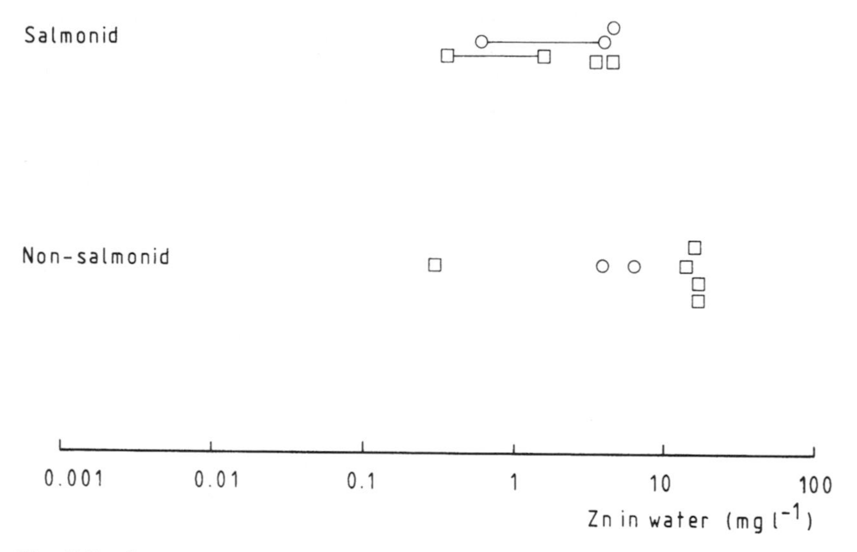

Fig. 7.7. Comparative effects of zinc on Salmonidae and other fish, expressed as LC_{50}. Details as for Fig. 7.5. Original figure compiled from values in the literature (see Appendix 2).

data from the control site Graham *et al.* (1986) assumed a mortality from injury alone to approach 20% after 12 days' exposure.

These experiments clearly have great value under certain circumstances; removing some of the uncertainty about the extent to which laboratory conditions resemble the field. Nonetheless they lack much of the control of environmental conditions possible within a laboratory and are themselves subject to some experimental effects, as demonstrated above (Graham *et al.*, 1986). For preliminary assessments of river conditions they may be perfectly adequate but most researchers have preferred the greater control of laboratory experiments. The literature on these is enormous and it is best considered in separate sections.

7.2.2.1 Reproduction and early life-stages

Many studies have stressed the major effects of heavy metals on reproduction and fecundity at concentrations well below the accepted LC_{50}. The 96-h TL (= tolerance limit)$_{50}$ for adults of the fathead minnow *Pimephales promelas* was between 0·60 and 0·98 mg litre^{-1} dissolved copper; however spawning was blocked at 0·18 mg litre^{-1} (Brungs *et al.*, 1976). Based on survival, growth, reproduction and hatchability of the eggs they proposed that the 'Maximum Acceptable Toxicant Concentration' (MATC) was between 0·066 and 0·118 mg litre^{-1} copper. Similarly for lead in *Salmo gairdneri*; the 96-h LC_{50} was between 1·32 and 1·47 mg litre^{-1} of the 'dissolved' metal in hard water but the MATC lay between 0·018 and 0·032 mg litre^{-1}. In soft water the LC_{50} was 1·17 mg litre^{-1} lead but if eyed eggs were exposed to the lead then their MATC was between 0.004 and 0.008 mg litre^{-1} and if the lead was introduced at hatching or to young fry then the MATC was between 0·007 and 0·015 mg litre^{-1} lead (Davies *et al.*, 1976). Speranza *et al.* (1977) exposed zebrafish *Brachydanio rerio* to the 'threshold concentration' (the highest sublethal dose) of zinc over nine days during the period in which the gametes were maturing. Not only did this delay spawning but the number of eggs produced was reduced to 38% of the controls and only 21% of these were viable compared with 90% of the controls. In addition the survival rate of the progeny was only 0·9% compared with 63% of the controls. These toxic effects are manifested differently in male and female fish; the LC_{50} of zinc to males of the vivparous fish *Lebistes reticulatus* was 300 mg litre^{-1} compared to 278 mg litre^{-1} in females. There were also different chronic effects in each sex; the males exhibited reductions in spermatogonia, spermatids and sperm whilst the females showed reductions in mature oocytes (Sehgal & Saxena, 1986).

The message from all of these studies is that reproduction is often a sensitive indicator of the chronic effects of heavy metals. In a recent review of the literature (Woltering, 1984) adult survival and growth were reduced in only 13% and 5 % respectively of all of the studies. In contrast fry survival was reduced in 57% of cases, fry growth in 36%, reproduction (number of eggs and/or spawns per female) in 30% and hatchability in 19% of cases. These indicate the importance of assessing effects on reproduction and the early life-history of fish in any study of the chronic effects of metals.

7.2.2.2 Behaviour

The effects of heavy metals on seven catagories of fish behaviour were discussed in a recent comprehensive review (Atchison *et al.*, 1987). These were: locomotion (including avoidance and attractance to metals), respiration, learning, reproduction, non-reproductive social interactions, feeding and predator avoidance. Reproduction has already been discussed (Section 7.2.2.1) and respiration will be covered in the next section along with the physiological aspects. Some of the remaining categories will be briefly covered here; for further details readers are referred to Atchison *et al.* (1987), in particular for their coverage of the features of experimental design which may affect results.

One obvious defensive reaction to elevated metal concentrations is to move away and try to find uncontaminated water nearby. There are several reports of this behaviour in the literature in response to concentrations as low as 0·0001 mg litre^{-1} copper and 0·0056 mg litre^{-1} zinc (Atchison *et al.*, 1987). These were all based in the laboratory; however there is some verification for these from field studies on the Northwest Miramichi River in New Brunswick, Canada (Sprague *et al.*, 1965). They defined an 'Incipient Lethal Level' (ILL) as the boundary between lethal and non-lethal concentrations. Concentrations higher than this tended to cause death within one day. For *Salmo salar* the ILLs for copper and zinc were 0·048 and 0·600 mg litre^{-1} respectively; however in the laboratory salmon parr avoided concentrations less than one tenth of these and when concentrations in the river rose to more than 0·35–0·43 of the ILL then abnormal downstream runs of adult salmon were observed (Sprague *et al.*, 1965).

Non-reproductive social interactions include the establishment and maintenance of territories and social hierarchies. Amongst populations of *Lepomis macrochirus* exposed to 0·034 mg litre^{-1} copper there was an increase in aggression in the dominant fish resulting in more threats and 'nips' in the population (Henry & Atchison, 1986). Similar results were obtained when *L. macrochirus* was exposed to zinc in the laboratory (Sparks *et al.*, 1972). Here pairs of fish kept in containers for periods of five days developed clear dominant–submissive relationships. This in itself was not unusual; however it became clear that the dominant fish survived exposure to the zinc for longer than the submissive fish. When a flowerpot was added to the aquarium to provide shelter there were reductions both in the number of aggressive encounters and in the difference in their response to the metal (Sparks *et al.*, 1972). Similarly, individual stone loaches *Noemacheilus barbatulus* appeared to lose the instinct to hide during daylight in response to high concentrations of cadmium in a laboratory study (Solbé & Flook, 1975).

7.2.2.3 Physiological effects and mechanisms of metal tolerance

There have been a great many studies on the effects of heavy metals on different aspects of fish physiology and any review here is necessarily selective with the aim of extracting common features. Part of the problem is the multifarious approaches to these studies — as an example different workers have studied both the effect of copper on food consumption and starvation (Collvin 1985*a,b*) and the effect of starvation on copper toxicity (Segner, 1987) — which makes the search for commonality more difficult.

The effect of copper toxicity on starvation and vice versa is a sensible place to start. At least part of this apparent ambiguity may stem from differences in the feeding patterns of the fish. Changes in concentrations of zinc in tissues of *Salmo trutta* took place only when the fish was feeding (O'Grady & Abdullah, 1985) emphasizing again the importance of the food supply as a source of heavy metals in the first place (see Chapter 5). There will also be features which relate to the overall metabolic state of the fish; however under these conditions starved fish may be more susceptible than fed specimens. After seven days exposure to 0·08 mg $litre^{-1}$ copper both fed and starved specimens of roach *Rutilus rutilus* had accumulated copper in the gills but only the starved fish accumulated the metal in the liver. If the fish were refed after seven days of starvation and contamination there was a decrease in the concentration of copper in the liver; however if starvation continued then the copper was retained (Segner, 1987). Within the liver there were decreases in the number of mitochondria and proliferation of endoplasmic reticulum; however there were no pathological lesions or copper-specific alterations. Starvation in this case acted to accentuate the effects of the copper. In contrast when perch *Perca fluviatilis* was exposed to 0·044 mg $liver^{-1}$ copper (0·15% of the 96-h LC_{50}) there was no change in the rate of food consumption but a decrease in the growth rate, accompanied by an increase in the starvation rate. Over fifteen days there was no acclimatization but a steady metabolic cost (Collvin, 1985*a*) attributable to a reduced food conversion efficiency because of an increase in the energetic costs associated with detoxification mechanisms (Collvin, 1985*b*). Both these and the study of Segner (1987) demonstrated a close involvement of copper toxicity with the metabolism of the fish; thus if there are marked energetic costs of sublethal exposure to heavy metals in fish fed maximal rations (Collvin, 1985*b*) then these effects will obviously be more pronounced when the fish has already been starved prior to exposure (Segner, 1987).

A second high metabolic cost results from the toxic effect of the metals on the gills with concomitant effects on respiration. In fact the effects of acute zinc toxicity to *Salmo gairdneri* were very similar to the effects of hypoxia (Burton *et al.*, 1972) and they speculated that a breakdown in gas exchange at the gills, either as a result of coagulation and precipitation of mucus or from cytological damage, was responsible. In tissues from fish subjected to high concentrations of zinc and the absence of oxygen there were both decreased concentrations of pyruvic acid, the usual end product of glucose metabolism, and elevated concentrations of lactic acid, the end product of glucose metabolism under oxygen-limiting conditions (Burton *et al.*, 1972). This hypoxia may be manifested in many diferent ways. Another study on *S. gairdneri* showed increases in heart rate (measured on an electrocardiogram), ventilation and coughing in response to treatment with 20 mg litre^{-1} zinc for 120 and 150 min. Some five to six days were subsequently required to recover normal rates for these processes and, in addition, the heart beat of the fish was closely coupled to ventilation during the first day after treatment which served to improve the conditions for effective oxygen transfer by the tissues (Hughes & Tort, 1985). In *Barbus conchonius* exposed to 0·047 mg litre^{-1} lead (20% of the 96-h LC_{50}) there was a decrease in the concentrations of erythrocytes and haemoglobin in the blood coupled with an anaemic appearance (Tewari *et al.*, 1987).

Recently some workers have disputed that heavy metal toxicity is simply a case of tissue hypoxia. When *Lepomis macrochirus* was exposed to the 96-h LC_{50} concentration of copper, 2·0 mg litre^{-1}, there were decreases in concentrations of adenosine triphosphate (ATP), adenosine diphosphate (ADP), total adenylates and energy charge in some tissues but no change in either the muscle or liver concentrations of lactic acid after 48 h indicating that tissue hypoxia alone was not the cause of these adenylate changes (Heath, 1984). At a copper concentration of 0·21 mg litre^{-1} there was an overall decrease in oxygen consumption by *L. macrochirus* over 32 days but no change in the in-vitro consumption by either the gills or the brain and only a slight increase in consumption by the liver (Felts & Heath, 1984). They suggested that the copper acted to decrease the oxygen consumption of the whole animal at a higher level of integration than those individual tissues. However it is also possible that the extent of the hypoxia at least in part varies between different taxonomic groups of fish.

A variety of other toxic effects of heavy metals have been observed although, as for plants, it is not always clear to what extent some of these are secondary consequences of more fundamental metabolic changes. One effect which may accentuate other toxic effects is decreased resistance to infection. Low concentrations of cadmium (but not chromium or copper) caused a slight reduction in the rate of antibody production by *Salmo gairdneri* in response to human red blood cells (Viale & Calamari, 1984) and 10 mg litre^{-1} zinc increased infection of a cell line of *S. gairdneri* by a virus which caused pancreatic necrosis (Hiller & Perlmutter, 1971). In this instance it was suggested that the zinc reduced the net negative charge of the cells and so enhanced adsorption of the virus by the cells. Other toxic effects which have been observed include decreased uptake of vital stains by cell lines of *Lepomis macrochirus* in response to nickel, copper, zinc and cadmium (Babich *et al.*, 1986), breakdown in the equilibrium mechanisms of *Semotilus margaritus* exposed to copper resulting in over-turning (Tsai, 1979) and decreased activity of δ-aminolevulinic acid dehydrase, an enzyme involved in haem synthesis, in two estuarine species (Jackim, 1973).

Some measure of acclimatization to chronic concentrations of heavy metals is possible; after 24 h exposure to 0·55 mg litre^{-1} copper *Salmo gairdneri* showed a 55% inhibition of sodium uptake and a 49% reduction in affinity for sodium with the net result of a decrease in its total sodium concentration of 12·5%. However after seven days the total sodium concentration was back to normal; although the rate of sodium uptake was still inhibited the fish had recovered by reducing its rate of sodium efflux (Laurén & McDonald, 1987*a*). After 24 h, accompanying biochemical changes included a 33% inhibition of gill Na^+–K^+–ATPase. Over the following 14 days this was compensated by a significant increase in the concentration of gill microsomal protein so that at the end of this period the total Na^+–K^+–ATPase activity per milligram of gill tissue was normal. The liver was the main site of copper accumulation and here, after seven days, there was an increase in the concentration of a sulphydryl-rich protein presumed to be a metallothionein (Laurén & McDonald, 1987*b*).

Metallothioneins have been frequently recorded in fish exposed to heavy metals, particularly in their liver. Typically they are low molecular weight proteins rich in sulphur-containing amino acids which are thought to bind and so detoxify certain heavy metals. In four-week-old chinook salmon *Oncorhynchus tshwytscha* only very low concentrations of metallothioneins were detected in response to 0·0025 mg litre^{-1} copper and 0·05 mg litre^{-1} zinc when the liver represented less than 1% of the total body weight; however by 21 weeks the concentration of metallothioneins was proportional to the ambient metal concentrations (Roch & McCarter, 1984). Induction of metallothioneins has been used by some workers as an indication of the 'no effect' concentration of heavy metals as the ambient concentration at which the metallothionein concentration in *Salmo gairdneri* was indistinguishable from the controls; based on linear regressions of hepatic metallothionein concentration versus ambient metal concentration the no effect concentrations were 0·002 mg litre^{-1} copper, 0·04 mg litre^{-1} zinc and less than 0·0002 mg litre^{-1} cadmium (Roch *et al.*, 1986).

7.2.2.4 Environmental factors mediating metal toxicity

As for plants and invertebrates the major factors affecting the toxicity of metals to fish are pH, hardness and the presence of chelating agents and, in turn, these are also factors which often affect metal accumulation. However this is not a simple cause-effect relationship and although uptake of copper by *Salmo gairdneri* was reduced by 50% at pH 5·0 compared with pH 7·0 the low pH acted additively to increase the apparent copper toxicity from between 0·200 and 0·400 mg litre^{-1} copper at pH 7·0 to between 0·025 and 0·100 mg litre^{-1} copper at pH 5·0 (Laurén & McDonald, 1986). Such effects are particularly pronounced in soft waters with a low buffering capacity and may affect all stages of the life cycle. For example, in an experiment designed to simulate the effects of acid precipitation a mixed 'spike' of seven metals, including nickel, copper, zinc and lead, was added to aquaria at 0·57 of the concentrations typically found in acidified water. This killed fry of the American Flagfish *Jordanella floridae* within six days as well as inhibiting spawning in adults and reducing the hatching of eggs (Hutchinson & Sprague, 1986). However in addition there are also records of increased toxicity at higher pH values. One such example based on factorial experiments tested the effects of both pH and hardness on zinc toxicity to *Pimephales promelas* using a continuous flow system. At each level of hardness zinc was more toxic at the highest pH values (Mount, 1966). At least part of the reason for this was attributed to the continuous flow system which tended to keep precipitated salts in suspension and caused increased mucus production by the fish. In another example the 96-h LC_{50} of metals to *Salmo gairdneri* were reduced from 0·066 mg $litre^{-1}$ copper and 0·671 mg $litre^{-1}$ zinc at pH 4·7 to 0·003 mg $litre^{-1}$ copper and 0·066 mg $litre^{-1}$ zinc at pH 7·0 (Cusimano *et al.*, 1986). In this instance the authors speculated that the reduced toxicity at low pH values was due to competition between the metal ions and hydrogen ions for binding sites. As both types of response seem to be found under certain circumstances, even in the same organism (compare Laurén & McDonald, 1986; and Cusimano *et al.*, 1986, for *S. gairdneri*) it is clearly difficult to generalize; however the interpretation of Campbell and Stokes (1985) with respect to type I and type II behaviour of metals (see Chapter 6) remains a useful starting point.

The effects of water hardness are easier to interpret as an increase in hardness generally decreases the toxicity of a metal (e.g. Mount, 1966). To a certain extent this is observed with most metals (Mance, 1987); however it tends to be less pronounced in those metals such as copper which bind strongly to ligands. No effect of increasing water hardness was observed, for example, for copper toxicity towards *Salmo gairdneri* (Laurén & McDonald, 1986). The theoretical binding constant of copper to glycine is 8·62 compared with 1·31 for calcium, 3·44 for magnesium and 5·52 for zinc. In juvenile *Salmo salar* there was some amelioration of zinc toxicity in the presence of magnesium but not in the presence of calcium but neither metal had any effect on copper toxicity (Zitko & Carson, 1976).

Other factors which affect metal toxicity in fish include the presence of chelating agents (decreased toxicity; Wilson, 1972), increased temperature (increased toxicity; Felts & Heath, 1984; Khangarot & Ray, 1987*a*) and synergism, addivity and antagonism with other metals (Table 7.1).

Chapter 8

Effects of Acid Mine Drainage On The Biota

8.1 ALGAE

8.1.1 Species composition

Some of the earliest studies on the effects of acid mine drainage on the biota were carried out in coal-mining areas of the Appelachians in the eastern USA where coal has been mined continuously since the 18th century (Bennett, 1969). In a survey of over 100 stream sites in Indiana and West Virginia no species of blue-green algae were found at a pH of less than 3·7 and only one below pH 7·0 (Lackey, 1938). The most acid-tolerant species were an unnamed species of *Chlamydomonas* and *Ulothrix zonata* (both Chlorophyceae), the diatom *Navicula* sp. (Bacillariophyceae), *Chromulina* sp. (Chrysophyceae) and the Euglenophyte *Euglena mutabilis*. A later study on lakes formed at the sites of strip mines in Ohio showed generally low concentrations of phytoplankton despite the lake pH never being recorded below 4·0 (Riley, 1960). Both the Chlorophyceae and the Bacillariophyceae were well represented; however species of *Euglena* were abundant at only one of these lakes. The most diverse phytoplankton flora was found at a 22-year-old lake with a near-neutral pH and comprised some 17 taxa of algae representing the Chlorophyceae, the Bacillariophyceae and the Dinophyceae.

A later study looked at a wide variety of aquatic habitats in the coal mining regions of West Virginia (Bennett, 1969). Unfortunately few details of precise pH and acidity ranges in which different organisms are found are given; however he does record a few species as being consistently abundant in mine entrances. These include *Ulothrix subtilis*, two diatoms (*Pinnularia braunii*, *Eunotia tenella*) and *Euglena mutabilis* along with three other species, (*Ulothrix* sp., *Frustulia rhomboides*, *Penium jenneri*) with a more seasonal distribution. *Euglena mutabilis* was the only alga of any significance at the two most acidic stations. *E. mutabilis* was abundant at the most acid reaches of another site in West Virginia, Roaring Creek (Warner, 1971). In this study a high diversity (greater than 33 species recorded) of periphyton was found where there was little or no acid pollution and the pH was greater than 4·9; however where the pH was less than 3·8 fewer than 20 species were recorded. He attributed the smothering effect of iron hydroxides to part of this but also noted that this abrupt decrease in species numbers took place at the point where the stream lost all of its natural bicarbonate buffering capacity (Fig. 8.1). Of the 20 species found in the more acid reaches ten (five Chlorophyceae, four diatoms and *E. mutabilis*) were particularly tolerant and of those *E. mutabilis* plus *Ulothrix tenerrima*, *Pinnularia termitina* and *Eunotia exigina* were present in large numbers only in the most acid reaches.

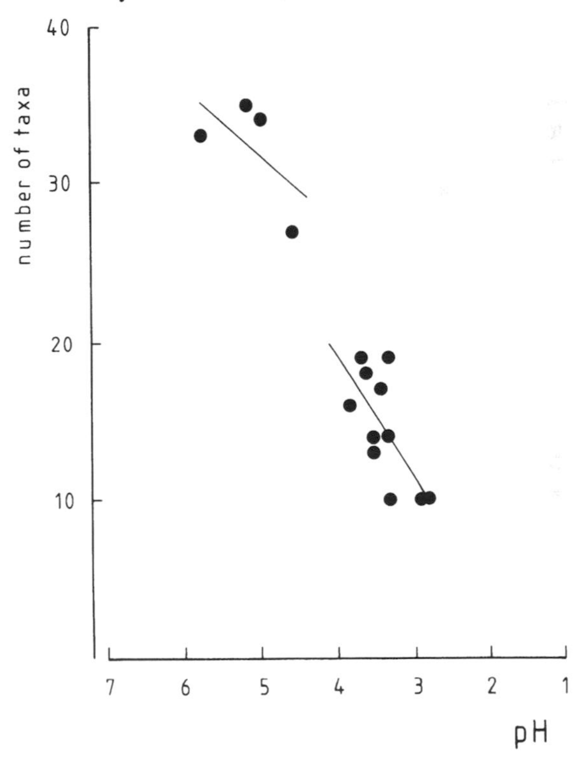

Fig. 8.1. Regression of diversity of periphyton against median pH at sites in Roaring Creek, West Virginia. (Redrawn from Warner (1971), *Ohio Journal of Science*, 71, 202–15 and reproduced with permission from the Ohio Academy of Science.)

In all of these early studies the same general effect was observed namely, a decrease in the number of species as the acidity increased, with only a few highly specialized organisms able to survive in the most highly acidic environments. Prime amongst these is *Euglena mutabilis*, by far and away the species most often recorded as abundant in acid mine drainage (Fig. 8.2). However it is not exclusively confined to this habitat with several records in *Sphagnum* peat (Hosiaisluoma, 1975; Pentecost, 1982), a very different type of acid habitat (Table 4.2). Nonetheless it is particularly well suited to the extreme acidity encountered in some acid mine drainages, with some growth detectable in the laboratory as low as pH 1·3 (Hargreaves & Whitton, 1976*a*). Unlike other members of the genus *E. mutabilis* does not possess a flagellum, instead moving by amoeboid-like contractions of its body which enables it to move easily in the absence of free water (Hosiaisluoma, 1975)

Euglena mutabilis was both the species found most frequently and often

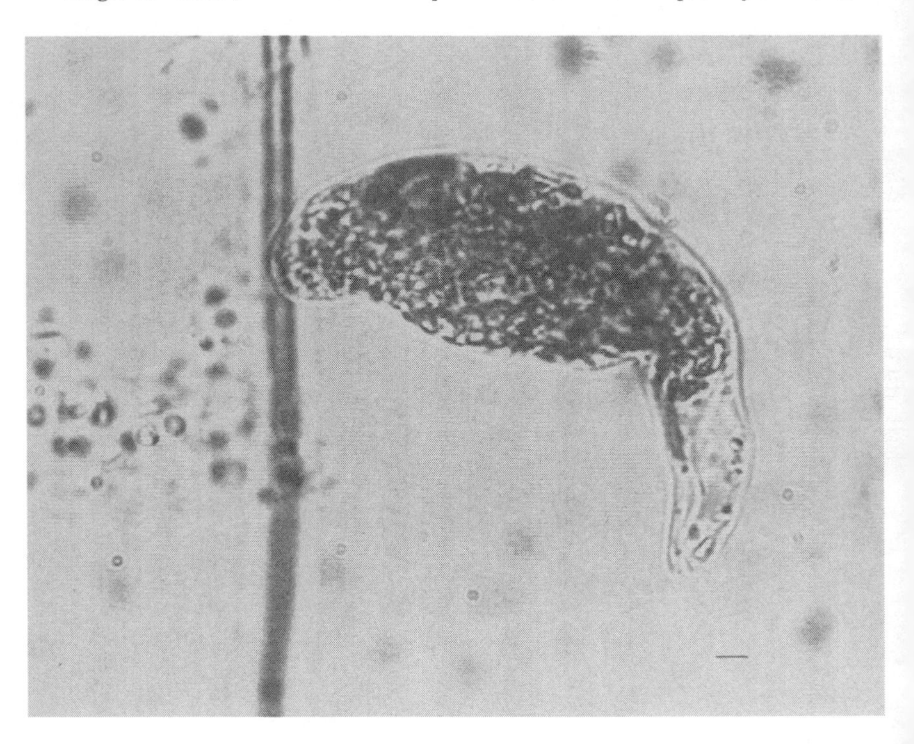

Fig. 8.2. Photomicrograph of *Euglena mutabilis* attached to a fungal hypha. Note the absence of a flagellum. In common with many other euglenophytes *E. mutabilis* lacks a rigid cell wall and hence its shape is constantly changing. Scale bar = 1 μm.

the most abundant species at a site in a survey of acid mine drainage sites in England with a pH of 3·0 or less (Hargreaves *et al.*, 1975). The next most abundant species included four diatoms (*Pinnularia acoricola, Nitzschia subcapitellata, N. elliptica* var. *alexandrina, Eunotia exigua*) and the Chrysophyte *Gloeochrysis turfosa* followed by two green algae; *Chlamydomonas applanata* var. *acidophila* (cf. *C. acidophila*) and *Hormidium rivulare*. Only four of the species were common to the earlier American studies and, conversely, five species recorded as widespread in England had not been recorded from the USA (Hargreaves *et al.*, 1975). The diatom *Pinnularia acoricola* illustrates this well with only scattered records from Java and South Africa until Carter (1972) found it in samples from an acid mine drainage in County Durham at pH 2·5. Subsequently Hargreaves *et al.* (1975) found it in 71% of the streams they visited — only *Euglena mutabilis* was more widespread — yet whilst Bennett (1969) and Warner (1971) found species of *Pinnularia*, neither recorded *P. acoricola*. Whether this indicates that it is not found in the New World or that different authors are calling the same organism different names is not clear. Certainly Denys (1984) draws attention to the wide variability in published descriptions which may confuse identification.

The high concentrations of heavy metals often associated with acid mine drainage may itself exert an influence on species composition; however it is often difficult to separate out the effects of the heavy metals from the effects of the acidity *per se*. Analysis of records of *Pinnularia acoricola* and *Eunotia exigua* in the field showed the former to be more tolerant of low pH with several records below pH 2·0; however *Eunotia exigua* was more likely to occur in the presence of higher concentrations of copper and zinc (Whitton & Diaz, 1981). For these two species at least some distinction between the effects of pH and heavy metals on their distributions can be made.

A more detailed analysis of the effects of the different factors was performed by Hancock (1973*a*, *b*) working on the Klip River, Southern Transvaal above and below acidic inputs from gold mines. Above the mines the water was dystrophic with a conductivity of between 23 and 230 $\mu S\,cm^{-1}$ and was dominated by Chlorophyceae (especially the Conjugales), diatoms, some blue-green algae and, at one site, the yellow-green alga *Tribonema* sp. Below the mines he distinguished two zones of pollution; the first, affected by sand dumps, was slightly acid in the dry season (pH 6·0), rising to 8·6 in the wet season and with a conductivity up to 2400 $\mu S\,cm^{-1}$. Here there were extensive swamp areas dominated by the emergent macrophytes *Phragmites* and *Typha*; however in shallow water where the cover of macrophytes was low there were also dense mats of *Tribonema* with diatoms, particularly *Achnanthes minutissima*, epiphytic on it (Hancock, 1973*a*). Where the acid water reached the river the pH dropped to between 2·5 and 3·0 and there was a substratum of shifting sand which prevented any algal growth, although there were communities of *Phragmites communis* along the margins and on sand spits. Downstream there was a clear recovery zone with a pH of between 3·9 (wet season) and 7·6 (dry season). Here Chlorophytes dominated, with *Stigeoclonium labricum, Ulothrix bipyrenoidosa* and *Oedogonium pisanum* epiphytic on the principal macrophyte, *Potamogeton*, with frequent production of zoospores by the *Stigeoclonium* and *Ulothrix*. Diatoms did not become dominant until the recovery had been completed and the pH was between 7·3 and 8·5 (Hancock, 1973*a*).

Hancock (1973*b*) was also able to use communities of diatoms to indicate different combinations of environmental conditions. The diatoms *Eunotia exigua, E. pectinalis* and *Frustulia rhomboides* were dominant in the dystrophic waters above the mines, indicating a preference for weak, organic acids, but were also represented below the main input of strong mineral acids; however *Pinnularia microstauron* var. *brébissonii* grew best in the stretches with mineral acids and not at all in the dystrophic reaches (Fig. 8.3). *Fragilaria familiaris* was intermediate in its preferences. Hancock (1973*b*) also recorded *Pinnularia acoricola* although he did not consider it widespread enough to act as an indicator.

Such specialized floras in habitats which are, in evolutionary terms, very recent is dependent largely upon the existence of naturally acidic environments. Acid hot springs such as those at Yellowstone National Park in Wyoming have been extensively studied (e.g. Brock, 1969); however the species which dominate these have to be not only acidophilic but also thermophilic. *Cyanidium caldarium*, a *Chlorella*-like alga with affinities with the blue-greens is the most characteristic organism of these habitats. The acidic environments resulting from mining activities tend to be rather cooler. Examples of naturally-occurring non-thermal highly acidic environments include the Kootenay Paint Pots in British Columbia studied by Wehr & Whitton (1983*a*) with a pH range of 3·1–4·7. They recorded fourteen species of algae (of which nine were diatoms, including *Pinnularia microstauron* — see above) although not *Euglena mutabilis* or *Chlamydomonas acidophila*, the two most distinctive species of acidic environments. *C. acidophila* was found in a volcanic lake in Japan, along with *Cyanidium caldarium* and *Pinnularia braunii* var. *amphicephala*. This lake, with pH values down to 0·9, contained active fumaroles at its base, responsible for

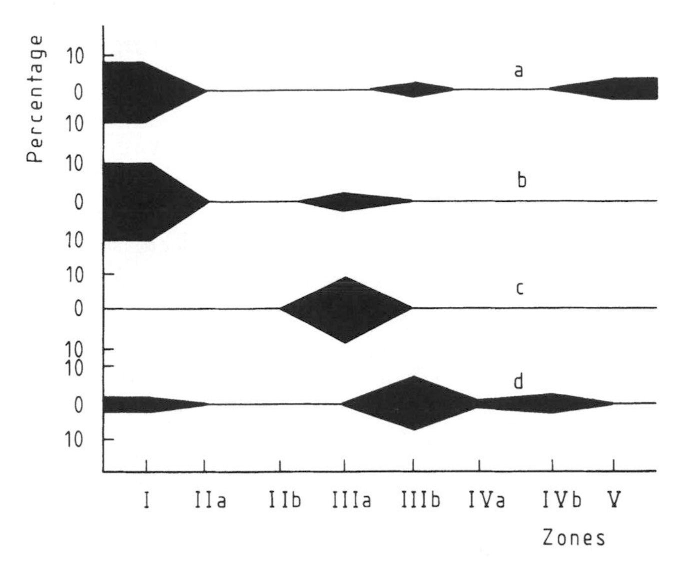

Fig. 8.3. Percentage distributions of the principal diatom species indicating acid water in the Klip River, Southern Transvaal. a, *Eunotia exigua* and *E. pectinalis*; b, *Frustulia rhomboides* and *F. rhomboides* var. *saxonica*; c, *Pinnularia microstauron* var. *brébissonii* and d, *Fragilaria familiaris*. pH of the different zones: Zone I, 5·9–7·5; Zone IIa, 6·3–8·6; Zone IIb, 2·5–3·0; Zone IIIa, 3·5–4·6; Zone IIIb, 4·7–7·7; IVa, 7·3–8·5; Zone V, 7·3–8·4. (Redrawn from Hancock (1973 *b*), *Hydrobiologia*, 42, 243–84 and reproduced with permission from Dr. W. Junk (Publishers).)

elevated concentrations of carbon dioxide but not, it appears, for significantly raised temperatures in the lake (Satake & Saijo, 1974).

A different type of natural acid habitat occurs at 'Smoking Hills' in the arctic region of the North West Territories, Canada, where spontaneous burning of bituminous shales releases large quantities of sulphur dioxide and sulphuric acid aerosols into the surrounding environment. In this respect it resembles an extreme form of 'acid rain' although the pH in affected tundra pools may fall as low as 1·8 (Sheath *et al.*, 1982). In comparison with the control pools (90 species) the total number of algal species found in pools close to the emissions was only 14 and the density of plankton in these pools was also much lower. In the control pools nine classes of algae were recorded with Cryptophyte *Chroomonas minuta* as the dominant organism; however only four classes were found in the affected pools. *Euglena mutabilis* dominated both the plankton and the periphyton of the most acidified pond (pH, 1·8; total acidity, 5800 mg litre^{-1} $CaCO_3$) followed by diatoms and *Chlamydomonas acidophila*. In the less acidic ponds (pH, 2·8 and 3·6; total acidity, 320 and 30 mg litre^{-1} $CaCO_3$) diatoms (*Eunotia arcus, E. glacialis, Nitzschia communis*) dominated, followed by *Chlamydomonas acidophila*; in these pools *Euglena mutabilis* formed only a minor part of the biomass.

It is reasonable to assume that all three of these natural non-thermal acid environments have been available for algal colonization for at least 1000 years; however at pH $< 4{\cdot}0$ none contain more than ten species, and the general effect is for the number of species to decrease along with the pH (Fig. 8.4). In other respects, however, the flora is similar whether or not the source of the acidity is natural or anthropogenic, particularly in the frequent abundance of *Euglena mutabilis* and *Chlamydomonas acidophila* especially at the lowest pH values. Diatoms, especially of the genera *Pinnularia, Eunotia* and *Nitzschia* are also common; presumably any species which are able to withstand the high acidity are able to capitalize on the high concentrations of silica which are often present.

Of the phyla of freshwater algae the Rhodophyceae are conspicuous by their absence from accounts of acid mine drainage and the Cyanophyceae are only rarely found at these pH values (Fig. 8.5). The lowest record here is an *Oscillatoria* sp. recorded by Warner (1971) at pH 2·8. In contrast, the

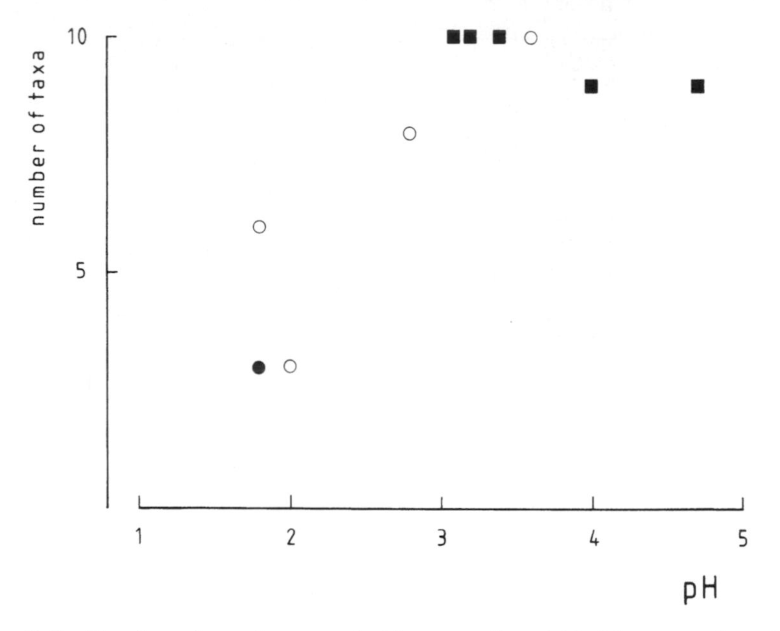

Fig. 8.4. Number of species recorded in naturally acid habitats as a function of pH. Original figure based on data in Satake and Saijo, 1974 (closed circles), Sheath *et al.*, 1982 (open circles) and Wehr and Whittton, 1983*a* (closed squares). Regression equation: $y = 1{\cdot}04 + 2{\cdot}28x$; $r^2 = 0{\cdot}57$; $p < 0.01$.

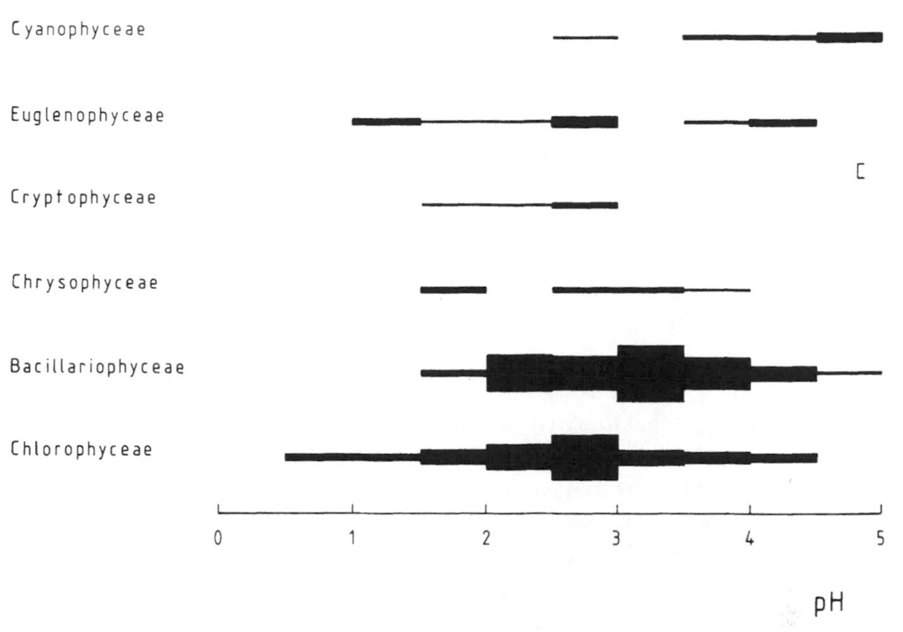

Fig. 8.5. Lowest pH at which different phyla of freshwater algae are recorded. Figure compiled from published records of acid mine drainage environments. Each record represents the lowest pH that a species is found at during each study and the apparent lack of records of some groups above pH 4·0 is an artefact of the search rather than a real phenomenon. Vertical bar = five records. See text for more details.

Chlorophyceae and Bacillariophyceae are well represented. The lowest records for Chlorophyceae are a *Chlorella* sp. and *Stichococcus* sp. recorded by Whitton & Diaz (1981) at pH 0·9. Other genera found below pH 2·0 include *Chlamydomonas, Characium* and *Desmidium* (Lackey, 1938; Hargreaves *et al.*, 1975; Whitton & Diaz, 1981). *Pinnularia acoricola*, found at pH 1·5 by Hargreaves *et al.* (1975), is the lowest record for the Bacillariophyceae on this graph, with a *Navicula* sp. at pH 1·8 (Lackey, 1938) and *Eunotia exigua* at pH 2·2 (Whitton & Diaz, 1981). Records for *Euglena mutabilis* in the field appear to extend down to pH 1·4 (Whitton & Diaz, 1981).

8.1.2 Experimental studies

There are a variety of difficulties facing any autotrophic organism which grows in highly acidic environments: concentrations of bicarbonate are either very low or non-existent, there are difficulties in maintaining a constant internal pH and there are often high concentrations of heavy metals. On the other hand, organisms which can overcome these problems may face relatively little competition for resources. Not surprisingly then acidophilic organisms have been the subject of much study both in the field and the laboratory.

Early experiments on *Euglena mutabilis* in the laboratory (Dach, 1943) indicated no growth at pH 0·9 but slight growth at pH 1·4. Maximum growth was observed between pH 4·0 and 4·8 and it subsequently declined, with only slight growth at pH 7·9 and none at pH 8·2. These figures at the lower end agree quite well with Hargreaves and Whitton (1976*a*) although maximum growth of their strain occurred at pH 2·0, close to the pH of the stream from which it was isolated; Fott and McCarthy (1964), however, record growth as low as pH 1·0. *Chlamydomonas acidophila* appears to exhibit similar behaviour with a pH optimum of about 2·5 and a rapid decline in growth above pH 6·0 (Hargreaves & Whitton, 1976*a*) although Cassin (1974) appears to have found the best growth at pH 5·0. Several of these papers also report enhanced growth in the presence of either the original substrate (Fott & McCarthy, 1964; Cassin, 1974) or a slight enrichment with membrane-filtered streamwater (Hargreaves & Whitton, 1976*a*). This may be a straightforward nutrient effect (Hargreaves & Whitton, 1976*a*) or the effect may be less clearcut — like the addition of soil extracts to microbial enrichment cultures. There are also reports of heterotrophy amongst these organisms; *Chlamydomonas acidophila* was able to grow in the dark with glucose as its sole energy source whilst a related, less-acidophilic species, *C. sphagnophila*, did not (Cassin, 1974). Similarly, both *Chlamydomonas acidophila* and *Euglena mutabilis* showed vigorous growth on glutamic acid (Fott & McCarthy, 1964). Facultative heterotrophy amongst the euglenophytes generally is well known. Under phosphorus-limiting conditions many algae are also able to produce phosphatases with which they are able to utilize phosphorus-containing organic compounds. A strain of *Chlamydomonas acidophila* was found to have two phosphatase enzymes capable of working under acid conditions which were present in cultures grown with and without phosphorus along with a third acid phosphatase produced as an adaptive response to phosphorus-starvation (Boavida & Heath, 1986).

Extensive experiments on the effects of nutrients on the growth of algae in acid lakes have been performed in Cheat Lake in West Virginia and Pennyslvania (DeCosta & Preston, 1980; Wilcox & DeCosta, 1982, 1984). This artificial lake has a number of strip mines and deep coal mines in its catchment and a typical pH of about 4·8. However, as a result of dilution of the original acid mine drainage the water is quite soft, with combined concentrations of magnesium and calcium of less than 25 mg $litre^{-1}$

(DeCosta & Preston, 1980). Its phytoplankton was dominated by the dinoflagellate *Peridinium inconspicuum* along with the chlorophytes *Botryoccus braunii* and *Chlorella* sp., the chrysophyte *Dinobryum divergens* and the euglenophyte *Trachleomonas volvocina* (Wilcox & DeCosta, 1984). In several respects these species are typical of many oligotrophic and dystrophic lakes. Laboratory assays of lake water using *Selanastrum capricornutum* showed that below pH 5·5 algal growth was limited both by the inorganic carbon supply and by phosphorus; however above pH 5·5 then phosphorus alone could stimulate growth (DeCosta & Preston, 1980). Phosphorus additions to polyethylene bags (1–1·75 m^3 capacity) suspended in the lake had a greater effect, possibly due to greater mixing of carbon dioxide with the water (DeCosta & Preston, 1980). However there was also a shift in dominance, so that the phytoplankton within the treated bags composed 90% of *Chlorella* sp. along with *Chlamydomonas* and *Scenedesmus quadricauda* (Wilcox & DeCosta, 1982).

Another question which puzzled physiologists for some time was how these organisms regulated their internal pH. If the interior pH was as low as the external pH chlorophyll would normally be degraded to phaeophytin unless there were special physiological adaptations; however maintenance of a near-normal pH would require energy-consuming ion pumps. Breakdown of chlorophyll in *Cyanidium caldarium* from acidic habitats occurred only when the membranes were disrupted by freezing and then thawing the cells in their medium, a procedure often used to disrupt membranes, indicating that in this species at least the pH inside the cell was probably higher than outside (Allen, 1959). More recently the effect of external pH on the internal pH of *Euglena mutabilis* and two species intolerant to high acidity, *Chlorella pyrenoidosa* and *Scenedesmus quadricauda*, was examined (Lane & Burris, 1981). The internal pH of the two intolerant species remained at about pH 7 regardless of the external pH; however in *Euglena mutabilis* there was some degree of regulation, with an internal pH of about 6·0 at an external pH of 3·0, rising to an internal pH of 8·0 at an external pH of 9·0. In *Chlorella pyrenoidosa* and *Euglena mutabilis* the rate of photosynthesis appeared to be independent of pH; however *Scenedesmus quadricauda* showed a rapid increase in photosynthesis above pH 5·5 as the bicarbonate supply increased (Lane & Burris, 1981). The low supply of inorganic carbon at acid pH values is a major problem although there is some evidence for *Chlorella saccharophila*, a species which can grow at pH 2·0, that there is a mechanism for concentrating carbon dioxide at the chloroplast envelope analagous to the C-4 mechanism of carbon fixation amongst higher plants (Beardall, 1981).

8.2 BRYOPHYTES AND HIGHER PLANTS

Several of the problems facing bryophytes and higher plants have already been mentioned with respect to algae. Even if a plant can overcome the acidity itself and the iron oxide precipitate it still has to be able to grow in the absence of bicarbonate. It is not surprising, then, to find that the two most abundant groups of macrophyte (*sensu lato*) in acid-mine drainage streams are the bryophytes, which use carbon dioxide rather than bicarbonate as their inorganic carbon source, and emergent angiosperms, which are not wholly dependent upon the inorganic carbon supply in the water.

In acid mine drainage streams in England (pH $< 3{\cdot}0$) both protonema and adult plants of two bryophytes (*Dicranella* sp. and *Drepanocladus fluitans*) and two emergent angiosperms, (*Juncus effusus* and *Typha latifolia*) were found in addition to algae (Hargreaves *et al.*, 1975). Fundamentally similar observations were made in Ohio (Lackey, 1938) with only one species of moss, *Catharinea* and *Typha latifolia* along with the pteridophytes *Isoetes* and *Equisetum arvense*. Where the stream had overflowed the bank vegetation was dead as well. The emergent angiosperm *Eleocharis acicularis* appeared to thrive at acid mine drainage sites in Pennyslvania (Ehrle, 1960).

It appears that, under conditions of environmental stress, bryophytes are often unable to grow beyond the protonema stage. In addition to the records above protonemal growths of the moss *Dicranella* have been observed at natural acid springs at the Kootenay Paint Pots (Wehr & Whitton, 1983*a*) and at zinc-rich seepages in the Old Lead Belt, Missouri (Whitton *et al.*, 1981) although it is still not clear what prevents them from developing into adult plants. It is also possible to confuse the branched, filamentous forms with some algae which makes interpretation of some records difficult. Generally the protonema can be distinguished by their oblique transverse cross walls (Fig. 8.6).

Downstream recovery of the macrophyte flora was observed below lignite mines in Denmark (Sand-Jensen & Rasmussen, 1978). At the highest sites, with a pH between 3·0 and 4·0 the flora was generally restricted to bryophytes (*Anisothecium vaginale, Polytrichum longisetum*) and emergent angiosperms (*Phragmites communis, Juncus bulbosus, J. effusus*); however further downstream, as the pH and alkalinity rose so submerged angiosperms such as *Elodea canadensis* and *Callitriche cophocarpa* were able to grow.

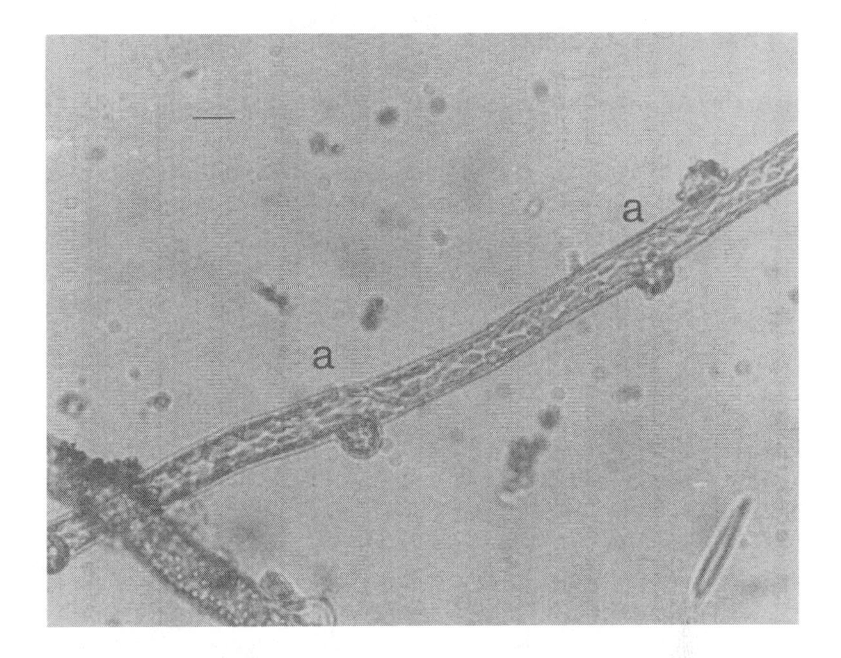

Fig. 8.6. Photomicrograph of protonema of the moss *Dicranella* sp. collected from a highly acidic habitat (pH 2·8). Moss protonema are often confused with filamentous green algae; however they characteristically possess obliquely transverse cross walls (a). Also obvious in this photograph are the buds from which the mature gametophyte develops. Scale bar = 50μm.

8.3 INVERTEBRATES

Interpretation of the effects of acid mine drainage on invertebrates is complicated by the variety of factors involved; in addition to the acidity itself there may be problems from high concentrations of suspended solids, from precipitation of iron(III) hydroxide and from elevated concentrations of heavy metals.

There is general agreement that under conditions of high acidity there is a drastic reduction in the number of species of invertebrates which are found; however in many of the studies those species which are able to tolerate the acidity thrive in the absence of competition and predators. In particular the Chironomidae are frequently found at high densities; examples include a strip-mine lake in Kansas (pH 3·2–3·6) dominated by *Tendipes* sp. (Stockinger & Hays, 1960), acid reaches of the Taff Bargoed,

South Wales (pH $> 3{\cdot}5$) dominated by *Conchapelopia pallidula* (Scullion & Edwards, 1980) and a stream in West Pennyslvania affected by acid mine drainage (pH, 2·6–3·0; total acidity, 456–1130 mg litre^{-1} $CaCO_3$) dominated by *Tendipes riparius* (Koryak *et al.*, 1972). Although the Chironomidae are clearly tolerant to acidity itself it is often the zone where the iron(III) hydroxides are deposited where the largest densities are observed (Koryak *et al.*, 1972; Greenfield & Ireland, 1978). Chironomidae comprised 99·5% of the fauna in an acid (pH 3·2) pool at a mining site in Quebec State contaminated with aluminium (2·6 mg litre^{-1}), copper (0·23 mg litre^{-1}) and zinc (4·9 mg litre^{-1}) in addition to iron (Wickham *et al.*, 1987).

Other taxa which appear to be common include the Tipulidae family of dipterans (Koryak *et al.*, 1972) and megalopterans, especially *Sialis* sp. (Dills & Roger, 1974; Stockinger & Hays, 1960; Warner, 1971; Roback & Richardson, 1969). Otherwise, the accounts are less consistent; some authors record Ephemeroptera (e.g. Lackey, 1938; Harrison, 1965; Carrithers & Bulow, 1973; Scullion & Edwards, 1980) whilst others note them as absent (e.g. Harrison, 1958; Stockinger & Hays, 1960). At least part of this ambiguity can be attributed to differences in pH and acidity of the receiving waters due to different amounts of oxidation of the pyrites in the first place and subsequent dilution. Based solely on pH (with all the limitations discussed in Chapter 4) there are few records of Oligochaeta and only scattered records of Insecta taxa below pH 4·0, whilst there are several records of Diptera and Megaloptera (especially *Sialis*) below pH 3·0 (Fig. 8.7). Crustaceans are usually conspicuous by their absence, presumably because of dissolution of their calcium-carbonate containing exoskeletons; records of *Gammarus* in two streams in Ohio as low as pH 2·2 (Lackey, 1938) appear to be almost unique and may represent drift from upstream. An eloquent demonstration of the effect of low pH on the number of crustacean species, albeit in moorland pools rather than acid mine drainage, is given by Fryer (1980). He showed a decline in the total number of species as the pH dropped, with a lower limit of about pH 3·0 (Fig. 8.8). The relationship was particularly pronounced below pH 5·0; above this other factors (diversity of habitats, calcium concentration) presumably had a greater effect on species numbers.

As for plants so there was an abrupt decrease in the number of invertebrate taxa recorded in Roaring Creek, West Virginia at about pH 4·2, the point at which all the natural bicarbonate buffering of the stream was lost (Fig. 8.9; Warner, 1971). This is also quite noticeable in Fig. 8.7. The implication is that below this regulation of the organisms pH becomes

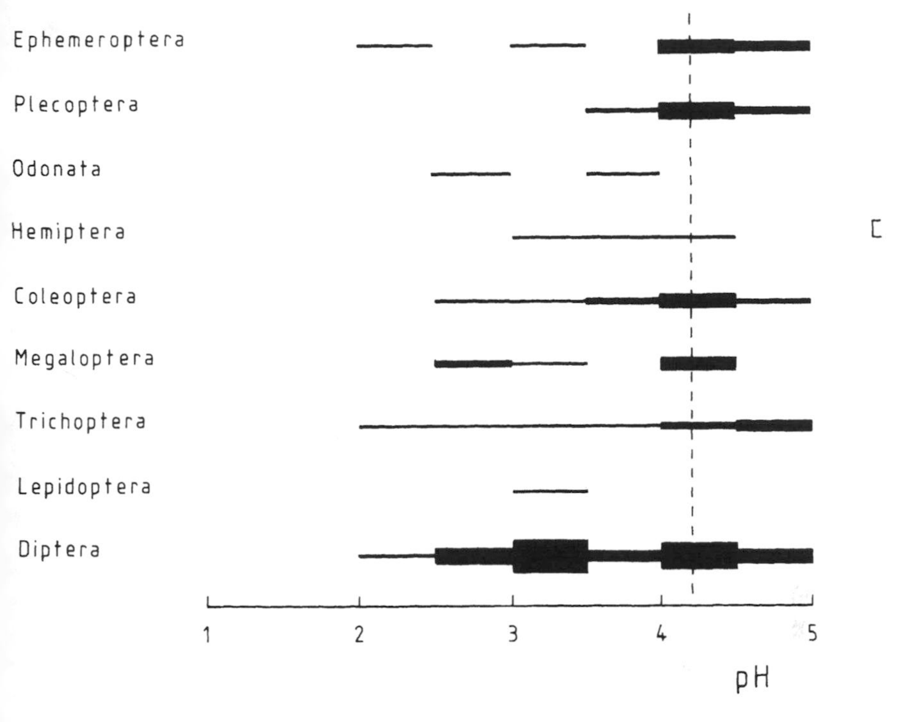

Fig. 8.7. Lowest pH at which different phyla of freshwater invertebrates are found. Figure compiled from published records of acid mine drainage environments. Each record represents the lowest pH that a species is found in a study and the apparent lack of records of some groups above pH 4·0 is an artefact of the search not a real phenomenon. Dashed line at pH 4·2 represents the point below which there is no bicarbonate alkalinity. Vertical bar = five records. See text for more details.

much more difficult and consequently only species with specialized adaptations can survive. One suggestion (Mackay & Kersey, 1985) is that 'shredder' species evolved an ability to tolerate local acidity within patches of decaying organic matter which then enabled them to colonize acidic habitats on a larger scale. This accounts for the widespread distribution of shredders in acidic softwater streams in Ontario despite the generally slower rates of decomposition of organic matter at low pH (Mackay & Kersey, 1985; Chamier, 1987). However there do not appear to be any species which 'indicate' acid mine drainage in the way that *Euglena mutabilis*, amongst the plants, does. The conclusion of Harrison (1965) was that the fauna in these streams was composed of fairly widespread species adapted to the acid conditions.

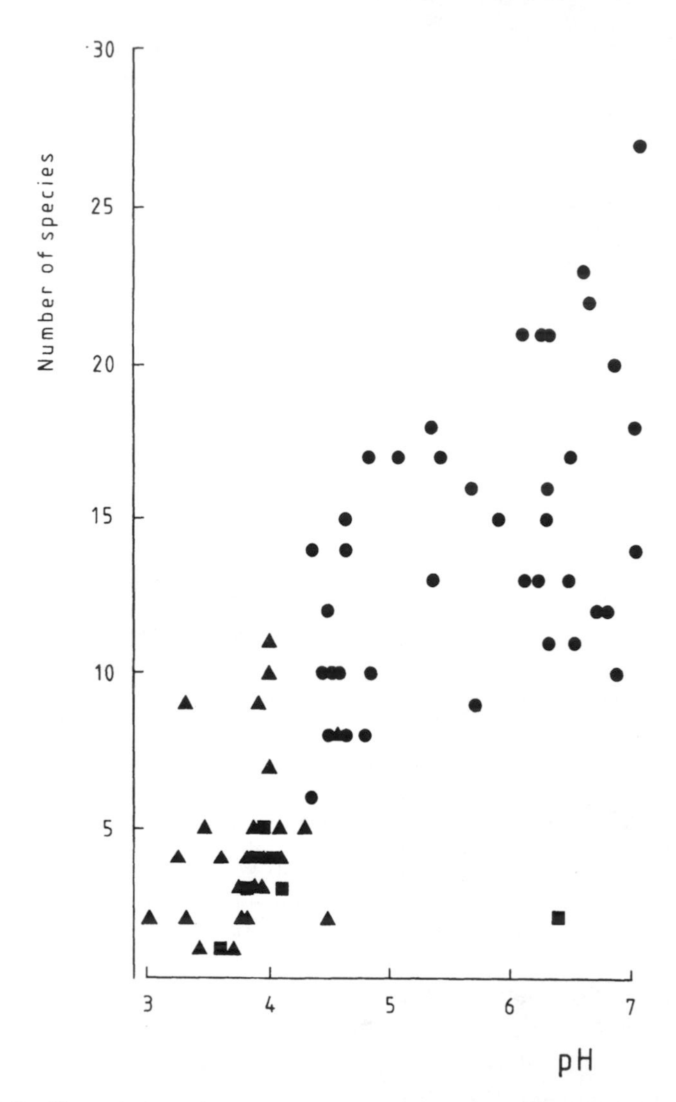

Fig. 8.8. The relationship between the number of species of crustaceans to pH in three moorland areas. Two Pennine sites, each with four species and a pH in the vicinity of 4, are omitted to avoid congestion. Samples are from Rhum (circles), Yorkshire Pennines (triangles) and north-east Yorkshire (squares). (Redrawn from Fryer (1980), *Freshwater Biology*, 10, 41–5 and reproduced with permission from Blackwell Scientific Publications.)

At very low pH values iron(III) hydroxide tends to remain in solution; however as the pH rises this is precipitated out and creates a different set

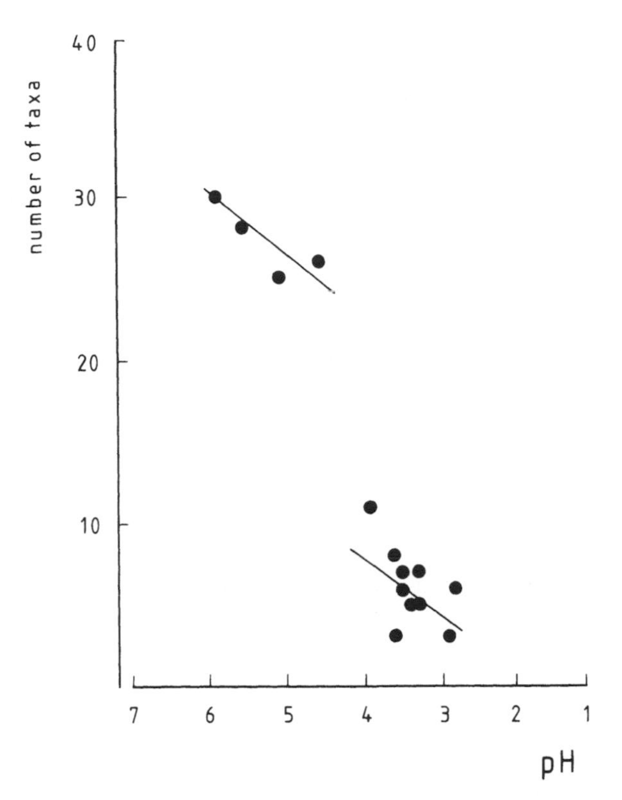

Fig. 8.9. Regression of diversity of benthic invertebrates against median pH at sites in Roaring Creek, West Virginia. (Redrawn from Warner (1971), *Ohio Journal of Science*, 71, 202–15 and reproduced with permission from the Ohio Academy of Science.)

of conditions for the benthos. In streams in West Pennsylvania riffles encountered in the region of neutralization (pH 3·3–3·8) had a greater number of species than the most acid reaches (pH 2·6–3·0), but their total biomass was much lower; the precipitated iron(III) hydroxide created more 'niches' for the benthos but created additional problems for any organisms which wanted to exploit them (Koryak *et al.*, 1972). Dipterans remained the dominant group, although the Tendipedids which were dominant at the highest site were joined here by Tipulidae, Sarcophagidae and Psychodidae, and a few ephemeropteran and coleopteran larvae began to appear. The first oligochaetes and plecopterans appeared at pH 3·8 (Koryak *et al.*, 1972). Similarly, over 200 organisms per square metre were observed in a reach of the Snake River, Colorado with a pH between 3·5 and 4·0 (McNight & Feder, 1984); however at a downstream site where the pH was higher (5·0–6·3) and where there was a covering of iron(III) hydroxide on the stream bed, this was drastically reduced and, again, those species which did survive may have represented drift from upstream sites (McNight & Feder, 1984).

One study which was particularly successful at separating the different effects of mine drainage was performed in the Taff Bargoed in South Wales (Scullion & Edwards, 1980). In this river there were distinct zones affected by acidic drainage, suspended solids and iron(III) hydroxide, all from different coal industry-derived effluents. For each there were initial decreases in abundance followed by zones of recovery downstream; however there were differences in the way in which each pollutant exerted its effect. In the uppermost reaches affected by acidic drainage (pH $> 3{\cdot}4$) the fauna was restricted to chironomids (mainly *Conchapelopia pallidula*) and the ephemeropteran *Baetis rhodani*. Downstream the Oligochaeta, Plecoptera and Trichoptera returned; however before the recovery was complete an input from a drift mine increased the concentrations of suspended solids. This reduced both the abundance of the fauna and the total number of species with the Ephemeroptera, Plecoptera and Trichoptera particularly sensitive, although certain taxa (such as *B. rhodani*) were more tolerant than others. In contrast Diptera (especially Chironomidae) and Oligochaeta thrived. Chironomidae and Oligochaeta along with *B. rhodani* also thrived below the iron(III) hydroxide-rich input. Scullion and Edwards (1980), like Harrison (1965) concluded that no species or taxa were particularly indicative of acid pollution; however it is clear from their results that some separation is possible between the effects of acid *per se* and associated factors such as suspended solids and iron(III) hydroxide. In particular the oligochaetes appeared to tolerate the physical pollutants but not the acidity.

The relative tolerances of groups of aquatic invertebrates are not necessarily borne out by laboratory experiments. In one case where the tolerances of larvae and nymphs of nine species were tested the two trichopterans were the most tolerant, with a 30-d TL_{50} of 2·45 and 3·38, followed by two plecopterans, two Odonata and an ephemoropteran (Bell, 1971). Very few trichopterans are recorded in the field at pH values as low as these. Successful emergence of the adult insects was more sensitive to pH than survival of the larvae and nymphs; the species followed a very similar order with 50% emergence by the most tolerant trichopteran at pH 4·0. As for heavy metals, then, toxic effects cannot be interpreted solely in terms of the response of the larvae.

8.4 FISH

In the reaches most polluted by acid mine drainage there are frequently no fish recorded. Several of the studies discussed above also considered fish; for example no fish were recorded in reaches with a lowest pH of 4·2 in the west fork of the Obey River, Tennessee (Carrithers & Bulow, 1973) nor in acid streams below gold and coal mining areas in South Africa (pH $> 2·9$). Warner (1971) found fish only where the pH was greater than 4·9 and Huckabee (1975) found reaches with a pH between 4·5 and 5·9 devoid of fish. On the other hand a strip-mine lake in Kansas (pH 3·2–3·6) was reputed to contain green sunfish (*Lepomis cyanellus*) and carp (*Cyprinus carpio*; Stockinger and Hays, 1960). This variety is great enough to suggest that fish distribution under these conditions is subject to more factors than pH alone.

Unfortunately, most pH measurements are based on spot readings and may miss short-term but important fluctuations. For example, intermittent discharges of acid water from a lignite mine were reported to be responsible for the death of trout in the near-neutral Lake Søby in Denmark which reduced the pH to 4·3 (Dahl, 1963). Similar effects have been shown in 'melt-water shocks' where snow containing high concentrations of sulphuric and nitric acid from acid precipitation thaws and the resulting drop in pH in streams may kill fish (Leivestad & Muniz, 1976).

High concentrations of suspended solids are frequently associated with acid mine drainage. Experiments using caged fish in the River Irwell, Lancashire showed these to have significant effects including production of excess mucus and the formation of deposits on the gills (Greenfield & Ireland, 1978). Parsons (1952) noted coagulation of the albumen in cells of the gills and a decreased permeability of cell membranes in response to acidity leading to suffocation of the fish. Associated with this were increases in the amounts of yawning and coughing behaviour in largemouth bass *Micropterus salmoides* observed when the pH was artificially lowered to 4·3 (Orsatti & Colgan, 1987). Related to this are findings by Lloyd and Jordan (1964) of a breakdown in the bicarbonate alkalinity in the blood of *Salmo gairdneri* at low ambient pH values. Increased yawning and coughing have been suggested as behavioural responses to hypoxia in fish subjected to elevated concentrations of heavy metals (Chapter 7). Alternatively, observed effects may be consequences of the death of food organisms (Parsons, 1952); however, *M. salmoides* also exhibited a decrease in the amount of time it spent feeding and instead spent more time hovering in the water column (Orsatti & Colgan, 1987).

As before, toxic effects may be exacerbated at early life stages and there may be chronic effects which reduce reproductive fitness (Cochran, 1987).

8.5 CONCLUDING REMARKS

Although in Chapter 4 the fundamental differences between different types of acidic habitats was discussed it is clear that in practice the difference between the low and high acidity is not clear cut. A mine effluent which conforms to the classic definition of acid mine drainage, may, once diluted, show some of the characteristics of the acidic, softwater environments caused by acid precipitation in the freshwater environment. From the point of view of a biologist concerned with monitoring these two are part of a continuum. On the other hand, in a base-rich area this overlap may never occur. To include all of the literature about the low pH, low acidity environments characteristic of regions affected by acid rain would have made this task impossible; a succinct summary is given by Kinsman (1984).

What the characteristic biota of streams receiving acid mine drainage are depends largely upon the extent of oxidation of pyrites in the first place. Where the pH is very low (i.e. $< 3{\cdot}0$) then both flora and fauna are very restricted; however whilst the flora is characterized by a few highly specialized organisms not found elsewhere (e.g. *Chlamydomonas acidophila, Euglena mutabilis*) the consensus in the literature is that the fauna comprises otherwise widespread organisms which happen to be tolerant to acid conditions. Also, based on the literature reviewed here it would appear that the flora is more tolerant to highly acidic conditions than the fauna (compare Figs. 8.5 and 8.7). In contrast, where the effects of acidity are secondary to the physical effects of precipitation of iron(III) hydroxide and suspended solids then it appears that a few animal taxa are the best equipped to survive.

In other words, there is no clear guide or advice for monitoring acid mine drainage; where the few 'indicator' algal taxa are not present high densities of Chironomidae larvae in the absence of known sources of organic pollution, along with an absence of submerged angiosperm species and the orange-brown precipitates of iron(III) hydroxide appear to be the best clues. Despite limitations, measurement of pH remains the quickest chemical confirmation of the nature of effects.

Chapter 9

Case Study: Placer Gold Mining

The very word 'gold' conjures up very different pictures to the other minerals discussed in this book. Countless Hollywood pictures have inscribed the images of the 1849 California gold rush firmly on our minds not as deep mines but as the 'Old Timer' painstakingly panning for the gold in the river. Whilst much of the world's supply now comes from deep mines, 'placer' (pronounced 'plass-er') mining, the modern equivalent of 'placer' deposits themselves are not confined to the surface; the Witwatersrand deposits in South Africa, the world's single largest source of gold, are comprised of fossil placers, although this is mined through deep mines. Some two-thirds of the world's known gold is in modern or fossil placers, with the Witwatersrand deposit comprising about 58% of this (Bache, 1987). Some of the environmental effects associated with the Witwatersrand deposit were discussed in Chapter 8. As surface placer mixing is so different in the type of environmental effects that it generates it is presented here as a contrast to these more conventional mining techniques. In particular thc modern placer mixing areas of Alaska and Yukon Territory in Canada have been the subject of a number of recent studies.

The continued existence of this type of mining is partly an acknowledgement of the high price of gold and partly a result of its great resistance to weathering and tendency to remain in its native state. Analysis of stream sediments remains a valuable technique for locating new sources of metals (Thornton, 1975); however it is only economically viable to work the sediments to extract a few metals, notably gold, platinum and tin. They are released as their host rocks are broken down and are transported by physical and chemical means in surface waters until redeposited. As a result of their high density this is likely to happen before many other minerals and creates relatively concentrated beds of gold in river gravels.

Traditionally this was separated from the sediment by crude flotation methods, the heavier gold sinking whilst the other minerals were washed away.

Modern techniques of placer mining are, not surprisingly, a long way removed from panning for gold. In regions of modern placer mining the deposits are stripped of vegetation and more recent deposits over the placer deposit. The deposits are then dredged using massive diggers mounted on barges and earth-moving equipment is used to shift it to the washing plants. The net effect is considerable disruption of the surrounding area as well as direct environmental effects on the streams. A significant part of the effect is attributed to turbidity in the water as a result of runoff and washwater contaminated by silt and clay in the deposits; however the disturbance caused when minerals are uncovered and exposed to weathering during mining operations can also release heavy metals.

Most attention has been concentrated on the effects of suspended solids. Turbidity is conventionally measured using a nephalometer, an instrument which measures the scattering of light by particles in suspension and standard 'nephalometric turbidity units' (NTU) calibrated against suspensions of polymers have been defined (American Public Health Association, 1981). These have provided a basis for legislation on the maximum amount of suspended solids in a stream above which stream life begins to suffer. In Alaska this is currently 25 NTU, and above mines this criterion is easily met, with typical values of 0·27 NTU and 0·7 mg litre^{-1} suspended solids (Pain, 1987). Below the mines, however, the figures for both are several orders of magnitude higher. Before mining began Ptarmigan Creek in Alaska had typical values of 2 NTU and 111 mg litre^{-1}; however after just a couple of days of mining values of 2800 NTU and 1126 mg litre^{-1} suspended solids were recorded (Bjerklie & LaPerriere, 1985). These problems are intensified if mines lack settling ponds or where several mines are concentrated in a catchment; in Birch Creek, Alaska (with about ten active mines) it is estimated that 277 tonnes of sediment per day were washed downstream whilst in nearby Faith Creek (four active mines) this figure was 29 tonnes per day (Bjerklie & LaPerriere, 1985). The effects of placer mining can extend far downstream of the sites of the mines; levels of turbidity and suspended solids 1000 times higher than the head of the stream have been recorded 92 km downstream of the last mine on Porcupine Creek, Alaska (Pain, 1987).

The release of suspended sediment has effects more far reaching than turbidity alone. Increased sedimentation results in the interfilling of interstices between gravels which has major ecological effects and, by silting the channel, can have marked effects on catchment hydrology. Normally, the streams would have acted as groundwater discharge points and there would be a close relationship between the level of the water table and the surface water level; however in mined streams the deposited sediment can act as a barrier to free movement to the stream. In streams below mines in Birch Creek, Alaska the water table was considerably lower than the surface water level, leaving the stream 'perched' above the water table and with lower concentrations of dissolved oxygen than streams in unmined catchments (Bjerklie & LaPerriere, 1985).

In order to put the effects of the suspended material into their correct ecological perspective the concentrations of heavy metals released by placer mining must also be considered. In the regions mined in Alaska copper, zinc, arsenic, mercury and lead were associated with the gold deposits and the water downstream of mines frequently contained elevated concentrations. Analyses of water collected below mines, however, revealed that most of this was associated with the particulate fraction (LaPerriere *et al.*, 1985). In Birch Creek in August 1983, for example, the total concentration of lead was 0·0362 mg litre^{-1} whilst the 0·45 μm filtrable fraction contained less than 0·0001 mg litre^{-1} (LaPerriere *et al.*, 1985). Similar relationships were observed for other metals and the filtrable concentrations, in general, were too low to exert strong effects on the biota. The metals in the particulate fraction, however, were presumably eventually transferred to the stream sediments and made subject to a range of physical, chemical and biological processes over time. Nonetheless, it appears that, in the short term, the effects of heavy metals are secondary to the effects of the silt and turbidity.

From the point of view of photosynthetic organisms the turbidity of the water has a major effect on the amount of photosynthetically active radiation (PAR) available to them. In a stream below a placer mine (5·2–170 NTU) PAR never rose above 120 μmol photon m^{-2}s^{-1} during the study period in August whilst in a nearby stream in an unmined catchment it was usually much higher than this. Nonetheless, because of the high latitude (> 65°N) even here it never rose much above 350 μmol photon m^{-2}s^{-1} in August. The effect of placer mining, therefore, can only be to decrease productivity in an already unproductive region. In addition, there was a positive correlation between the vertical light extinction coefficient and turbidity (r^2 = 0·99; Van Nieuwenhuyse & LaPerriere, 1986).

As a result, there were generally lower levels of primary production at sites downstream of mines than at control sites with no detectable primary production at all at a site on Birch Creek below the most heavily mined catchment. Similarly, there was no colonization of artificial substrates in Birch Creek, although substrates at sites below unmined and moderately mined catchments were colonized, predominantly by pennate diatoms. About twelve species were found at control sites, dominated by *Synedra ulna, Hannaea acrus, Achnanthes* spp. and *Diatoma tenue* var. *elongatum*; however at the intermediate site only two species (*Syndra acus, Eunotia* sp.) were found (Van Nieuwenhuyse & LaPerriere, 1986). It is not clear what relation these bear to the species which colonize natural substrates in the river; an environment where silt is constantly being deposited and shifted will present particular problems for any algal species. Best adapted to survive are those which are capable of some movement e.g. gliding. The blue-green alga *Plectonema* is an example of a species which can colonize shifting banks of sediment.

The ratio of photosynthesis to respiration is often used as a guide to the predominant trophic levels. In many tundra streams, where there are no trees to act as sources of organic matter for heterotrophic organisms this ratio is high; however in the streams below placer mines this tended towards unity (Van Nieuwenhuyse & LaPerriere, 1986). There appear to be few published studies on the benthic fauna of streams below placer mines; however a study of the drift fauna in Minto Creek in the Canadian Yukon above and below a tributary with an active placer mine has been made (Birtwell *et al.*, 1984). The highest density of invertebrates was found in the tributary itself (> 12 organisms m^{-3}), although the number of species was lower here than at the other sites. Chironomids, in particular, were dominant below the mine. They contrasted their results with studies on other types of mining with similar effects. Communities composed of small numbers of few species have been found below china clay mines for example and these were dominated by burrowing or tube-building organisms such as Tubificidae, Naididae and Chironomidae (Nuttall & Bielby, 1973). To this extent these communities differ from communities below other types of mining where chironomids are often both dominant and numerically abundant.

The highest trophic level in these food chains is usually dominated by the Arctic grayling, *Thymallus arcticus*, although several other species have also been found. In the presence of heavy placer mining these disappear. The suspended sediments have effects on different aspects of their biology including behaviour, reproduction and physiology. The deposited sediment removes refugia and the interstitial places in which they lay their eggs as well as reducing the amounts of oxygen for eggs and fry. By reducing the numbers of invertebrates it also affects their feeding habits. There is quite a lot of controversy on the precise effects that silt has on the fish; certainly many species have to be used to intermittent periods of natural high turbidity and migrating salmon have been observed to swim through glacial silt of the same order as that produced by mining activities although they chose clear tributaries in which to spawn (Smith, 1940). However in areas of California where there was widespread placer gold mining these clear tributaries were quite rare and the redds dug by spawning Chinook salmon (*Oncorhynchus tschawytscha*) were so close together as to overlap (Smith, 1940). Deposition of silt soon after the eggs had been laid appeared to delay the emergence of Silver salmon (*O. kisutch*) and reduce the yield of fry (Shaw & Maga, 1943).

Physiological effects of high concentrations of suspended solids on *Thymallus arcticus* collected from the field include abnormal development of gills, such as swelling and increase in epithelial cell size (termed hypertrophy), abnormal increases in epithelial cell numbers (hyperplasia) and clubbing (thickening of the distal ends of secondary lamellae; Birtwell *et al.*, 1984). Under laboratory conditions, however, suspended solid concentrations up to 250 g $litre^{-1}$ for four days or 1 g $litre^{-1}$ for six weeks failed to induce similar symptoms or to affect respiratory capabilities (McLeay *et al.*, 1987). There were reductions in feeding activity, greater downstream displacement and colour changes in fish exposed to mining silt in the laboratory (McLeay *et al.*, 1987); however under natural conditions these effects may be reduced by avoidance behaviour by the fish.

Some measure of control is possible using settling ponds to trap the sediment; however this alone may not be enough to satisfy water quality criteria and chemical flocculants may also be necessary. Alternatively the wastewater may be recycled through the sluices used to separate the gold from the gravels. There is also a degree of self-purification and miners such as Karl Hanneman (interviewed in Pain, 1987) argue that streams below sites mined 50 years ago are now healthy. With time the stream itself can wash away much of the offending silt. However, in the final analysis there will inevitably be a trade-off between economics and the environment to determine the best practicable option.

There has been a tremendous amount of work on the effects of heavy metals and acidity on the aquatic ecosystem at the expense, perhaps, of associated effects such as turbidity and suspended solids. Focussing on a problem such as placer mining allows these to come to the forefront and their effects to be separated from the effects of heavy metals and acidity *per se*. Placer mining may represent an extreme case of turbidity and suspended solids; however it is clear that it too can have severe effects on stream ecosystems. Most striking, perhaps, is the absence of photosynthetic organisms from the most affected sites; this is in marked contrast to the sites with the highest concentrations of metals and acidity. The intention of the chapter, however, is not just to draw attention to this but rather to divert attention away from a straightforward adherence to the (presumed) cause/effect relationships between heavy metals and the aquatic biota and towards a more holistic approach to environmental problems caused by mining generally.

Chapter 10

Conclusions

Before closing, some general comments on the 'state of the art' may be relevant. The amount of literature on the effects of heavy metals on freshwater ecosystems is enormous; however it is not immediately clear how much better off we are as a result. The variety of water bodies into which mine-derived pollution is discharged is such that the number of possible responses is almost infinite and the number of generalizations which can be made is quite small. Treat mine polllution as forms of environmental stress and a few common features are apparent: the decline in species numbers, accompanied by, perhaps, increases in the density of the few tolerant species able to thrive with the reduced competition and gradual changes in community composition along environmental gradients as the pollution is ameliorated are, perhaps, the three most obvious.

Under these circumstances the value of the 'case-study'-type of approach is obvious; however experimental studies have proved invaluable in extending the above generalization to include the effect of other environmental factors such as pH and calcium concentration on toxicity and to identify stages in the life cycle of organisms which are particularly susceptible to the pollutant. Concepts such as the LC_{50} have clear value for making comparisons between organisms, life-stages and different combinations of environmental conditions; however many authors have stressed the importance of chronic toxicity effects over the type of acute effect exemplified by the LC_{50}.

Despite the plethora of experimental studies there are still gaps in our knowledge, even of such basic concepts as the amelioration of metal toxicity by calcium. There is still a great need for workers to go beyond experimental observations and grapple with the physiological mechanisms behind these. More significantly, perhaps, there appear to be few critical comparisons of the effect of calcium on different metals in the way that Campbell and Stokes (1985) were able to do for pH. The relationship between accumulation and toxicity needs to be more rigorously researched and the causal links established.

Perhaps more basic yet is the concentration of studies, both field and laboratory, in the temperate regions of Europe, North America and Australia. By contrast, much of the mining for these metals now takes place in tropical and sub-tropical regions. This is no different to other fields of the environmental sciences and, historically, the science that we study is a product of Western civilization. Where some measure of reductionism has been applied in studies in temperate regions some of the general findings may be transferrable; however where the parts have to be combined to form a new whole there will be gaps, as indeed there are when such processes are applied to temperate ecosystems, exacerbated by the differences in the intensities of fundamental processes in tropical ecosystems.

Perhaps it is most appropriate to end upon a note of hope and rather than look forward to what still is to be done (something all scientists like to do), look back at how far the subject has come in the last 60 years or so since pioneers such as Kathleen Carpenter made their first studies. Such are the levels of awareness of the general public and understanding of environmental scientists towards environmental pollution that, in the developed world at least, industry can no longer get away with indiscriminate pollution. More significant, perhaps, than this negative approach is the 'pre-emptive' approach of the more responsible industrial companies; identifying the problems in advance and taking the initiative to develop appropriate responses. Such enterprise can only be applauded.

Appendix 1

Toxicity Tables — Plants

Part of the original aim for this book was to present published data on the toxicity of heavy metals to plants and animals in tabular form. Geoffrey Mance pre-empted this in his recent book (1987) which contains comprehensive tables on the toxicity of metals to invertebrates and fish. To include similar tables here would merely be repetition; however by tabulating data on the toxicity of metals to plants the two books may, perhaps, be seen as complements. In Appendix 2 data on the effects of metals on invertebrates and fish as used in the original figures in Chapter 7 are tabulated for quick reference and comparisons.

The concept of the LC_{50} as a standard toxicity test is nowhere near as established for algae and aquatic plants as it is for other groups, which makes comparison of studies more difficult. Some workers use LC_{50} or EC_{50} ('effective' concentration to reduce growth by 50%), whilst others use concentrations which completely inhibit growth or other parameters. Nonetheless it should be possible from these tables to get some indication of the relative tolerances to different metals.

Unless otherwise stated, hardness and alkalinity are presented as mg litre^{-1} $CaCO_3$. A lack of information in the tables reflects the absence of these details from the original reference.

Table 1

Species	*Test conditions*	*Metal salt added*	*Response criteria*	*Metal concentration (mg litre⁻¹)*	*Reference*
NICKEL					
Algae					
Scenedesmus acutiformis (laboratory strain) *Scenedesmus acutiformis* (lake strain)	8-d test, 28°C, illumination 16:8 light: dark	$NiSO_4$	Growth stopped Growth decreased	0·5 1·5	Stokes *et al.* (1973)
Scenedesmus (laboratory strain) *Scenedesmus* (lake strain)	8-d test, 28°C illumination 16:8 light: dark	$NiSO_4$	Growth 75% reduced Growth stopped Growth 40% reduced Growth stopped	0·2 1·5 0·7 1·5	Stokes (1975*a*)
Scenedesmus acutiformis var. *alternans*	6-d test, 28°C, illumination 16:8 light: dark		Growth 47% reduced Growth 82% reduced	1·0 3·0	Stokes (1975*b*)
Navicula pelliculosa	14-d test, 20°C, illumination 16:8 light: dark	$Ni(NO_3)_2$	Reduction in growth and biomass Toxic threshold	0·1 0–0·1	Fezy *et al.* (1979)
Anacystis nidularis (wild type) *Anacystis nidularis* (tolerant strain)	32°C, continuous light 120 μmol photon $m^{-2}s^{-1}$		Inhibition of growth	0·16 1·3	Whitton and Shehata (1982)

Higher plants					
Elodea canadensis	24-h test 24°C	$NiCl_2$	50% reduction in O_2 evolution	723	Brown and Rattigan (1979)
	28-d test		50% plant damage	2·8	
Lemna minor	28-d test		50% plant damage	0·34	
Lemna minor (*valdiviana*)	3-wk test, 24°C illumination 16:8 light: dark	$Ni(NO_3)_2$	Reduction in growth	0·3	Hutchinson and Czyrska (1975)
			Growth prevented	1·0	
COPPER					
Algae					
Sporotetras pyriformis	6-d test, 20°C, continuous light	$CuSO_4$	'Just non-inhibitory'	0·5	Whitton (1970)
Ulothrix				0·29	
Microspora				0·3	
Stigeoclonium tenue				0·3	
Ulvella frequenis				2·28	
Gongrosira				0·7	
Cladophora glomerata				0·12	
Oedogonium				0·17	
Mougeotia				0·28	
Spirogyra				0·24	
Chlorella pyrenoidosa	20°C, continuous illumination	$CuSO_4$	24-h delay in ordinary growth	0·001	Steemann-Nielson and Wium-Anderson (1970)
			48-h delay in ordinary growth	0·005	
Nitzschia palea	20°C, illumination 12:12 light:dark		72-h delay in ordinary growth	0·006	
Scenedesmus quadricauda	20°C, continuous illumination	$CuSO_4$	Lethal dose	0·8–1·0	Gibson (1972)
Anabaena flos-aquae			Lethal dose	0·1–0·25	

Table 1 — contd.

Species	*Test conditions*	*Metal salt added*	*Response criteria*	*Metal concentration (mg litre^{-1})*	*Reference*
Scenedesmus (laboratory strain) / *Scenedesmus* (lake strain)	8-d test, 28°C, illumination 16:8 light:dark	$CuSO_4$	25% growth decrease	0·05	Stokes *et al.* (1973)
			Toxic	0·1	
			25% growth decrease	1·0	
			50% growth decrease	2·0	
Selanastrum capricornutum	24°C, continuous illumination	$CuCl_2$	Initiates growth inhibition	0·05	Bartlett *et al.* (1974)
			Complete inhibition	0·09	
			Algicidal	0·3	
Scenedesmus acutiformis var. *alternans*	28°C, 6-d test illumination 16:8 light:-dark	$CuSO_4$	Initiates growth inhibition	0·1	Stokes (1975*a*)
			Severe growth inhibition	1·0	
	10-d test		Initiates growth inhibition	0·1	
			Severe growth inhibition	1·0	
Aphanizomenon / *Anabaena*	16–25°C	$CuSO_4$	Nitrogen fixation	0·01	Elder and Horne (1978)
			Completely repressed	0·01	
Selanastrum capricornutum	3-w test, 22°C		50% reduction in total algal cell volume.	0·085	Christensen *et al.* (1979)

Spirogyra, *Oedogenium*, *Microspora*, *Mougeotia*, *Ulothrix*, *Draparnaldia*	14-d test, continuous illumination	$CuSO_4$	Decrease in growth	0·065	Francke and Hillebrand (1980)
			100% lethality	6·5	
Aphanizomenon flos-aquae	5–7-d test, 17–22°C, continuous illumination	$CuSO_4$	20% reduction in C fixation	0·01	Wurtsbaugh and Home (1982)
			75% reduction in C fixation	0·03	
			Significant reduction of N_2 fixation	0·02	
			Total reduction of N_2 fixation	0·03	
			Chlorophyll loss complete	0·03	
Oscillatoria		$CuSO_4$	Growth inhibition	0·5	Arora and Gupta (1983)
			Beginning of separation discs	1·0	
			Well defined separation discs	10	
Microspora pachyderma	20°C, dim illumination	$CuSO_4$	Median log, MLD. (minimum lethal dose)	0·62	Foster (1982*b*)
Hormidium subtile				0·5	
Protoderma viride				0·5	
Pleurococcus vulgaris				1·25	
Chlamydamonas heterogama				1·0	
C. debaryana				0·5	
C. terricola				1·19	
C. asymmetrica				1·25	
C. applanata				0·5	
Gloeococcus mucosus				1·25	
Asterococcus limneticus				1·25	

Table 1 — contd.

Species	*Test conditions*	*Metal salt added*	*Response criteria*	*Metal concentration (mg litre^{-1})*	*Reference*
Hypnomonas				0·75	Foster (1982*b*) — contd.
Trochiscia granulata				0·5	
Chlorella vulgaris				1·0	
C. kessleri				0·5	
C. fusca				2·0	
C. protothecoides				0·5	
Selanastrum minutum				0·5	
Scenedesmus obliquus				0·5	
S. longus				1·0	
S. armatus				1·0	
S. acutiformis				1·0	
Spondylosium pygmaeum				0·75	
Anacystis nidulans (wild type)	32°C, continuous illumination 120 μmol photon $m^{-2}s^{-1}$		Inhibition to growth	0·15	Whitton and Shehata (1982)
Anacystis nidulans (tolerant strain)			Inhibition to growth	0·55	
Cladophora glomerata	Continuous light, 4000 lux, 20°C, 4-h incubation	$CuSO_4$	ED_{50} (median effective dose) based on $^{14}CO_2$ fixation	0·036	Robinson and Hawkes (1986)
Navicula incerta	16:8 h light:dark, 5500 lux, 19°C, modified Bristol medium	$CuCl_2$	96-h EC_{50} based on cell counts	0·010	Rachlin *et al.* (1983)

Scenedesmus quadricauda	90 μmol photon $m^{-2}s^{-1}$ continuous light, 20°C, Fraquil medium, shaken	$CuSO_4$	5-d EC_{50}	0·006	Petersen (1982)
Synechococcus sp.	1000 lux, 25°C, BG11 medium, shaken	$CuCl_2$	Minimum inhibitory concentration	0·0005	Lorenz and Krumbein (1984)
Anabaena sp.				0·002	
Higher plants					
Elodea nuttallii	14-d test, 21°C, Illumination 14:10 light: dark	$CuSO_4$	50% dead in 14 days	0·1	Marquenie-van der Werff and Pruyt (1982)
Elodea canadensis	24°C 24-h test	$CuSO_4$	Oxygen evolution 50% reduced	0·15	Brown and Rattigan (1979)
	28-d test		Phytotoxicity, 50% damaged	3·1	
Lemna minor	28-d test		Phytotoxicity, 50% damaged	0·13	
Lemna paucicostata	7-d test, 25°C, continuous illumination	$CuSO_4$	Frond multiplication rate reduced to 30%	0·3	Nasu and Kugimoto (1981)
Lemna minor	3-w test, 24°C, illumination 16:8 light: dark	$CuSO_4$	Leaf production reduced	0·05	Hutchinson and Czyrska (1975)
			Leaf production ceased	0·3	
ZINC					
Algae					
Selanastrum capricornutum	Algal assay procedure bottle test, continuous light, 20°C.	$ZnCl_2$	Initial inhibition of growth	0·03	Bartlett *et al.* (1974)
			Complete inhibition of growth	0·12	
			'Algicidal'	0·70	

Table 1 — contd.

Species	*Test conditions*	*Metal salt added*	*Response criteria*	*Metal concentration (mg litre^{-1})*	*Reference*
Sporotetras pyriformis	Continuous light, 600 lux, 20°C, modified Chu 10 medium, shaken	$ZnSO_4$	'Just non-inhibitory'	0·7	Whitton (1970)
Ulothrix				1·2–2·0	
Microspora				2·4–3·0	
Stigeoclonium				0·7	
Ulvella frequens				0·7	
Gongrosira				2·2	
Cladophora glomerata				0·08	
Dedogonium				0·08–0·14	
Mougeotia				0·5–3·0	
Spirogyra				0·18–4·0	
Chlorella vulgaris	2000 lux, 14:10 light:dark, modified Chu 10 medium, standing	$ZnCl_2$	50% survival	25·0	Rai *et al.* (1981)
Nitzschia linearis	Soft water media	$ZnCl_2$	5-d TL_m (50%) reduction in number of cells produced)	4·3	Patrick *et al.* (1968)
Cladophora glomerata	Continuous light, 4000 lux, 15°C, continuous flow culture	$ZnSO_4$	Reduction of specific growth by 50%	0·07	Robinson and Hawkes (1986)
Navicula incerta	16:8 light:dark, 5500 lux, 19°C, modified Bristol medium	$ZnCl_2$	96-h EC_{50}, based on cell counts	10·10	Rachlin *et al.* (1983)
Scenedesmus quadricauda	Continuous light, 90 μmol photon m^{-2} s^{-1}, Fraquil medium, shaken	$ZnSO_4$	5-d EC_{50}	0·6	Petersen (1982)

Nodularia incerta	Modified Hughes medium	ZnO	Maximum concentration permitting growth	2·0	Gopal *et al.* (1975)
Anacystis nidulans	500-600 lux, 28 ± 2°C, modified Hughes medium		Maximum tolerance level	1·5	Rana and Kumar (1974)
Oscillatoria sp.				1·0	
Arthrospira jenneri				1·5	
Plectonema boryanum				30·0	
Nodularia spumigena				2·0	
Anabaena doliolum				1·0	
Fisherella mucicola				1·0	
Chlorella vulgaris				25·0	
Scenedesmus sp.				30·0	
Anabaena spiroides			Inhibition of growth	0·005	Kostyayev *et al.* (1981)
			Complete suppression of growth	0·05	
			50% inhibition of nitrogen fixation	0·0025	
Chroococcus paris	1076 lux, 26°C, BG11 medium, shaken	$ZnSO_4$	Lowest concentration which showed detectable toxicity	> 0·4	Les and Walker (1984)
			Almost complete inhibition	2	
Spirulina platensis	Continuous light, 3000 lux, Zarrcuks medium	$ZnSO_4$	5-d LC_{50}	5	Kotangale *et al.* (1984)
Synechococcus sp.	1000 lux, 25°C, BG11 medium, shaken	$ZnSO_4$	Minimum inhibitory concentration	0·01	Lorenz and Krumbein (1984)
Anabaena sp.				0·001	

Table 1 — contd.

Species	*Test conditions*	*Metal salt added*	*Response criteria*	*Metal concentration (mg litre^{-1})*	*Reference*
Chlorella vulgaris	5000 lux, 16:8 light:dark, 15·5 ± 1°C, modified Bristol medium	$ZnSO_4$	96-h LC_{50}	2·4	Rachlin and Farran (1974)
Chlorella vulgaris	2000 lux, 14:10 light:dark, 24 ± 1°C, modified Chu 10 medium	$ZnCl_2$	50% 'survival' at three weeks	12	Rai *et al.* (1981)
Higher plants					
Lemna paucicostata	6000 lux, 25 ± 1°C, media supplied with 1% sucrose	$ZnSO_4$	Inhibition of frond multiplication	1 (soft water) 10 (hard water)	Nasu and Kugimoto (1981)
LEAD					
Algae					
Anabaena sp.	16:8 light:dark, 20°C, 4-h incubation	$Pb(NO_3)_2$	ED_{50} (median effective dose) based on $^{14}CO_2$ fixation	15	Malanchuk and Gruendling (1973)
Chlamydomonas reinhardti				17	
Navicula pelliculosa				17	
Cosmarium botrytis				5	
Sporotetras pyriformis	Continuous light, 6000 lux, 20°C modified Chu 10 medium, shaken 6-d incubation	$PbCl_2$	'Just non-inhibitory'	15	Whitton (1970)
Ulothris				27–35	
Microspora				40–48	
Stigeoclonium tenue				20	
Ulvella frequens				16	

Gongrosira				19	
Cladophora glomerata				2	
Oedogonium				12–32	
Mougeotia				19–42	
Spirogyra				14–31	
Cladophora glomerata	Continuous light, 4000 lux, 15°C, continuous flow culture	$PbCl_2$	Reduction of specific growth rate by 50%	1·03	Robinson and Hawkes (1986)
Lyngbya sp. (wild type)	Modified Chu 10 medium	$Pb(NO_3)_2$	Complete inhibition of growth	0·01	Bagchi *et al.* (1985)
(tolerant strain)				0·03	
Navicula incerta	5500 lux, 16:8 light:dark, 19°C, modified Bristol medium	$PbCl_2$	96-h EC_{50} based on cell counts	10·96	Rachlin *et al.* (1983)
Selanastrum capricornutum	EPA standard algal assay medium, 22°C, 3 w-test		50% reduction in growth	0·14	Christensen *et al.* (1979)
Anabaena sp.	1300 lux, 16:8 light:dark, 20 ± 1°C, BG11 medium	$Pb(NO_3)_2$	Complete inhibition		Laube *et al.* (1980)
Synechococcus sp.	1000 lux, 25°C, BG11 medium, shaken	$Pb(CH_3\text{-}COO)_2$	Minimum inhibitory concentration	> 0·05	Lorenz and Krumbein (1984)
Anabaena sp.				> 0·05	
Hormotila blennista	3000–4000 lux, 22 ± 1°C,	$PbCl_2$	50% inhibition of growth	0·08	Monahan (1973)

Appendix 2

Sources of Published Data Used in Original Figures

Table A1
Fig. 4.1. Relationship between pH and acidity

pH	*Total acidity* (*mg litre*$^{-1}$ $CaCO_3$)	*Reference*
Field measurements		
2·6	1130	Koryak *et al.* (1972)
3·0	456	
3·8	146	
4·8	21	Roback and Richardson (1969)
5·3	13	
3·2	44	
7·0	11	
8·1	8·3	Carrithers and Bulow (1973)
4·4	300	
5·0	82	
7·0	17	
7·4	9·9	
8·0	6·9	
8·2	7·8	
3·3	104	Warner (1971)
3·3	139	
3·3	130	
3·8	46	
3·5	77	
3·6	75	
4·2	91	
4·6	10	
5·7	2	
2·9	408	
2·8	509	
3·5	100	
4·9	5	
5·2	6	
1·8	5800	Sheath *et al.* (1982)
2·0	6000	
2·8	320	
3·6	30	
Laboratory measurements		
1·0	647	Bell (1971)
2·0	393	
3·0	59	
4·0	42	
5·0	15	
6·0	7·6	
7·0	2·1	
7·8	1·5	

Table A2

Figs. 5.1 and 5.5. Concentrations of nickel measured in water (mg litre^{-1}) and plant (μg g^{-1}) and enrichment ratios (concentration in plant/concentration in water). Published values only. A = alga; B = bryophyte; P = pteridophyte; S = sporophyte (angiosperm)

Group	*Taxa*	*Concentration*		*Enrichment ratio*	*Reference*
		Water	*Plant*		
A	*Mougeotia*	0·15	240	1600	Trollope and Evans (1976)
A	*Tribonema*		160	1067	
A	*Tribonema*		260	1733	
A	*Tribonema*	0·10	290	2900	
A	*Coccomyxa*	0·12	150	1250	
A	*Zygnema*		700	5833	
A	*Oscillatoria*	0·12	1070	8917	
A	*Ulothrix*	2·20	300	136	
A	*Microspora*	2·94	110	37	
A	*Cladophora*	0·07	30	429	
A	*Spirogyra*		130	1857	
A	*Oedogonium*	0·06	70	1167	
A	*Cladophora*	0·12	100	833	
A	*Spirogyra*		90	750	
A	*Spirogyra*		30	250	
P	*Equisetum palustre* (roots)	0·042	193	4595	Hutchinson *et al.* (1975, 1976)
	(shoots)		79	1881	
S	*Nuphar vaginatum* (leaves)		47	1119	
	(petioles)		35	833	
	(peduncles)		5	119	
	(roots)		14	333	
S	*Potamogeton* sp. (leaves)		480	11428	
	(stems)		255	6071	

P	*Equisetum palustre* (roots)	0·003	48	16000
	(shoots)		18	6000
S	*Potamogeton* sp. (leaves)		50	16667
	(stems)		18	6000
P	*Equisetum palustre* (roots)	0·002	24	12000
	(shoots)		13	6500
S	*Nuphar variegatum* (leaves)		8	4000
	(shoots)		7	3500
	(peduncles)		3	1500
	(roots)		6	3000
S	*Potamogeton* sp. (leaves)		39	19500
	(stems)		7	7500

Table A3

Figs. 5.2 and 5.6. Concentrations of copper measured in water (mg $litre^{-1}$) and plant ($\mu g\ g^{-1}$) and enrichment ratios (concentration in plant/concentration in water). Published values only. A = alga; B = bryophyte; P = pteridophyte; S = sporophyte (angiosperm)

Group	*Taxa*	*Concentration*		*Enrichment ratio*	*Reference*
		Water	*Plant*		
S	*Ceratophyllum*	0·0042	16	3960	Fayed and Abd-El-Shafy (1985)
S	*Eichhornia*		22	5167	
S	*Panicum*		3	762	
S	*Eichhornia*	0·0048	90	18688	
S	*Panicum*		3318	691250	
S	*Eichhornia*	0·0045	20	4511	
S	*Panicum*		4	956	
S	*Eichhornia* (root)	0·0039	44	11308	
	(shoot)		20	5205	
S	*Panicum* (root)		48	12256	
	(stem)		4	974	
	(leaf)		6	1462	
S	*Eichhornia*	0·0051	22	4412	
B	*Hygrohypnum luridum*	0·0019	45	23684	Jones *et al.* (1985)
B	*Scapania undulata*	0·0045	160	35554	
B	*S. undulata*	0·0037	108	29189	
B	*Jungermannia vulcanicola*	0·006	8	1333	Satake *et al.* (1984)
B	*J. vulcanicola*	0·025	62	2480	
B	*J. vulcanicola*	0·17	23	135	
S	*Elodea nuttallii*	0·0254	438	17244	Ernst and Marquenie-van der Werff (1978)

S	*Spirodela polyrhiza*		40	1575	
S	*Ceratophyllum demersum*		33	1299	
S	*Elodea nuttallii*	0·0127	113	8898	
A	*Vaucheria*		13	1024	
S	*S. polyrhiza*	0·0064	13	2031	
S	*C. demersum*		33	5156	
A	*Vaucheria*		5	781	
S	*C. demersum*	0·0191	28	1466	
A	*Oedogonium*		20	1047	
S	*S. polyrhiza*	0·0064	39	6094	
S	*C. demersum*		42	6562	
S	*Nuphar luteum*		6	938	
A	*Oedogonium*		40	6250	
S	*E. nuttallii*	0·0191	68	3560	
S	*N. luteum*		17	890	
A	*Vaucheria*		13	681	
S	*N. luteum*	0·0191	23	1204	
A	*Vaucheria*		13	681	
S	*E. nuttallii*	0·0064	18	2812	
S	*S. polyrhiza*		11	1719	
S	*C. demersum*		14	2188	
S	*N. luteum*		6	938	
A	*Vaucheria*		4	625	
A	*Oedogonium*		6	938	
P	*Equisetum arvense* (shoot)	0·043	27	628	Ray and White (1979)
	(root)		181	4209	
P	*E. arvense* (shoot)	0·014	74	5286	
	(root)		348	24857	
P	*E. arvense* (shoot)	0·012	17	1417	
	(root)		43	3583	

Table A3 — contd.

Group	*Taxa*	*Concentration*		*Enrichment ratio*	*Reference*
		Water	*Plant*		
P	*E. arvense* (shoot)	0·001	3	3000	
	(root)		11	11000	
P	*E. arvense* (shoot)	0·001	6	6000	
	(root)		9	9000	
B	*Fontinalis squamosa*	0·60	105	175	McLean and Jones (1975)
B	*Scapania undulata*		202	337	
B	*S. undulata*	1·80	64	36	
A	*Mougeotia*	0·03	380	12667	Trollope and Evans (1976)
A	*Tribonema*		400	13333	
A	*Tribonema*		700	23333	
A	*Tribonema*	0·02	670	33500	
A	*Coccomyxa*	0·01	650	65000	
A	*Zygnema*		460	46000	
A	*Oscillatoria*	0·02	340	17000	
A	*Ulothrix*	0·05	480	9600	
A	*Microspora*	0·06	1020	17000	
A	*Cladophora*	0·03	60	2000	
A	*Spirogyra*		220	7333	
A	*Spirogyra*		290	9667	
A	*Oedogonium*	0·02	110	5500	
A	*Cladophora*	0·01	50	5000	
A	*Spirogyra*	0·02	230	11500	
A	*Spirogyra*		50	2500	
P	*Equisetum palustre* (roots)	0·009	88	9778	Hutchinson *et al.* (1975, 1976)
	(stems)		28	3111	

S	*Nuphar variegatum* (leaves)		18	2000
	(petioles)		23	2556
	(peduncles)		11	1222
	(roots)		20	2222
S	*Potamogeton* sp. (leaves)		81	9000
	(stem)		48	5333
P	*E. palustre* (roots)	0·003	18	6000
	(shoots)		6	2000
S	*Potamogeton* sp. (leaves)		18	6000
	(stems)		15	5000
P	*E. palustre* (roots)	0·002	26	13000
	(shoots)		7	3500
S	*N. variegatum* (leaves)		4	2000
	(shoots)		2	1000
	(peduncles)		2	1000
	(roots)		5	2500
S	*Potamogeton* sp. (leaves)		27	13500
	(stems)		15	7500

Table A4

Figs. 5.3 and 5.7. Concentrations of zinc measured in water (mg litre^{-1}) and plant (μg g^{-1}) and enrichment ratios (concentration in plant/concentration in water). Published values only. A = alga; B = bryophyte; P = pteridophyte; S = sporophyte (angiosperm)

Group	*Taxa*	*Concentration*		*Enrichment ratio*	*Reference*
		Water	*Plant*		
S	*Ceratophyllum*	0·0014	47·8	34143	Fayed and Abd-El Shafy (1985)
S	*Eichhornia*		52·5	37500	
S	*Panicum*		30·5	21786	
S	*Eichhornia*	0·0017	34·2	20118	
S	*Panicum*		194·8	114588	
S	*Eichhornia*	0·0115	64·6	5617	
S	*Panicum*		31·3	2722	
S	*Eichhornia* (root)	0·0121	105·8	8744	
	(shoot)		74·5	6157	
S	*Panicum* (root)		105·4	8711	
	(stem)		35·5	2934	
	(leaf)		39·9	3298	
S	*Eichhornia*	0·0066	68·4	10364	
B	*Hygrohypnum luridum*	0·35	690	1971	Jones *et al.* (1985)
B	*Scapania undulata*	0·14	1160	8286	
B	*S. undulata*	0·30	1480	4933	
B	*Jungermannia vulcanicola*	0·067	17	254	Satake *et al.* (1984)
B	*J. vulcanicola*	0·027	43	1592	
B	*J. vulcanicola*	0·034	14	412	
B	*Fontinalis squamosa*	0·024	97	4042	Say *et al.* (1981)
B	*F. squamosa*	0·032	197	6156	

B	*Rhynchostegium riparioides*		197	6156	
B	*F. squamosa*	0·52	2390	4596	
B	*R. riparioides*		2160	4154	
B	*R. riparioides*	0·23	1120	4870	
B	*F. squamosa*	0·25	1980	7920	
B	*R. riparioides*		2130	8520	
S	*Eleocharis obtusa*	0·894	1040	1163	Adams *et al.* (1980)
S	*Potamogeton robbinsii*	0·624	1250	2003	
S	*P. robbinsii*	0·0125	583	46640	
S	*P. pectinatus*		336	26880	
S	*P. crispus*		224	17920	
S	*Elodea canadensis*		818	65440	
A	*Cladophora* sp.		393	31440	
S	*P. robbinsii*	0·0311	376	12090	
S	*P. pectinatus*		327	10514	
S	*Ceratophyllum demersum*	0·0292	343	11747	
S	*C. demersum*	0·0312	374	11987	
S	*E. canadensis*		349	11186	
S	*C. demersum*	0·0356	377	10590	
S	*P. pectinatus*		272	7640	
P	*Equisetum arvense* (shoot)	0·339	551	1625	Ray and White (1979)
	(root)		841	2481	
P	*E. arvense* (shoot)	0·081	1017	12556	
	(root)		1618	19975	
P	*E. arvense* (shoot)	0·090	282	3133	
	(root)		331	3637	
P	*E. arvense* (shoot)	0·001	30	3000	
	(root)		65	65000	
P	*E. arvense* (shoot)	0·001	33	33000	
	(root)		41	41000	

Table A4 — contd.

Group	*Taxa*	*Concentration*		*Enrichment ratio*	*Reference*
		Water	*Plant*		
A	'Phytoplankton'	0·008	1310	163750	Cushing (1979)
B	*Fontinalis squamosa*	1·20	125	104	McLean and Jones (1975)
B	*Scapania undulata*		245	204	
B	*S. undulata*		1950	1625	
A	*Mougeotia*	34·1	44940	1318	Trollope and Evans (1976)
A	*Tribonema*		19930	584	
A	*Tribonema*		17610	516	
A	*Tribonema*	19·61	21110	1076	
A	*Tribonema*		17440	889	
A	*Coccomyxa*	11·44	19050	1665	
A	*Zygnema*		45890	4011	
A	*Oscillatoria*	1·96	1880	959	
A	*Ulothrix*	4·9	3560	726	
A	*Microspora*		9260	1890	
A	*Cladophora*	0·21	890	4238	
A	*Spirogyra*		1590	7571	
A	*Spirogyra*		1920	9143	
A	*Oedogonium*	0·08	120	1500	
A	*Cladophora*		970	12125	
A	*Spirogyra*	0·16	1090	6812	
A	*Spirogyra*	0·05	320	6400	

Table A5

Figs 5.4 and 5.8. Concentrations of lead measured in water (mg litre^{-1}) and plant (μg g^{-1}) and enrichment ratios (concentration in plant/concentration in water). Published values only. A = alga; B = bryophyte; P = pteridophyte; S = sporophyte (angiosperm)

Group	Taxa	Concentration		Enrichment ratio	Reference
		Water	Plant		
S	*Ceratophyllum*	0·0067	11·8	1 761	Fayed and Abd-El-Shafy (1985)
S	*Eichhornia*		12·2	1 821	
S	*Panicum*		11·4	1 702	
S	*Eichhornia*	0·0077	10·2	1 325	
S	*Panicum*		12·1	1 571	
S	*Eichhornia*	0·0095	6·5	684	
S	*Panicum*		4·8	505	
S	*Eichhornia* (root)	0·0090	5·2	578	
	(shoot)		1.7	189	
S	*Panicum* (root)		7·8	867	
	(stem)		6·1	678	
	(leaf)		14·8	1 644	
S	*Eichhornia*		13·2	1 467	
B	*Hygrohypnum luridum*	0·008	4 770	596 250	Jones *et al.* (1985)
B	*Scapania undulata*	0·18	49 400	274 444	
B	*S. undulata*	0·06	18 300	305 000	
B	*Fontinalis squamosa*	0·005	99	19 800	Say *et al.* (1984)
B	*F. squamosa*		53	10 600	
B	*Rhynchostegium riparioides*		59	11 800	
B	*R. riparioides*	0·004	36	9 000	
B	*R. riparioides*	0·005	10	2 000	
B	*F. squamosa*	0·006	29	4 833	

Table A5 — contd.

Group	*Taxa*	*Concentration*		*Enrichment ratio*	*Reference*
		Water	*Plant*		
B	*R. riparioides*		39	6 500	
S	*Phalaris* (shoot)	0·005	31	6 200	Welsh and Denny (1976)
	(root)		360	72 000	
S	*Potamogeton crispus* (shoot)		380	76 000	
	(root)		1150	230 000	
S	*Phalaris* (shoot)	0·002	38	19 000	
	(root)		138	69 000	
S	*Potamogeton crispus* (shoot)		100	50 000	
	(root)		138	69 000	
B	*Fontinalis squamosa*	0·08	10 800	135 000	McLean and Jones (1975)
B	*Scapania undulata*		14 825	1 853 125	
B	*S. undulata*	0·04	12 375	309 375	
A	*Mougeotia*	0·31	6 190	19 968	Trollope and Evans (1976)
A	*Tribonema*		5 450	17 581	
A	*Tribonema*		4 940	15 936	
A	*Tribonema*	1·24	3 680	2 968	
A	*Tribonema*		14 190	11 444	
A	*Coccomyxa*	0·1	3 230	32 300	
A	*Zygnema*		2 600	26 000	
A	*Oscillatoria*	0·1	580	5 800	
A	*Ulothrix*	0·31	2 380	7 677	
A	*Microspora*	2·91	2 160	742	
A	*Cladophora*	0·1	230	2 300	
A	*Spirogyra*		400	4 000	

A	*Spirogyra*	110	1 100
A	*Oedogonium*	60	600
A	*Cladophora*	90	900
A	*Spirogyra*	130	1 300
A	*Spirogyra*	40	400

Table A6

Fig. 7.1. The recovery of the invertebrate fauna in the River Rheidol, mid-Wales, following the cessation of lead mining

Year	*Number of species*		
	Total	*Trichoptera*	*Reference*
1920	14	0	Carpenter (1924)
1922	29	8	Carpenter (1924)
1931	100	17	Laurie and Jones (1938)
1937	68	—[a]	Laurie and Jones (1938)
1949	130	18	Jones (1949)

[a] No data available.

Table A7

Figs 7.2–7.4. Effect of nickel, copper, zinc and lead on aquatic invertebrates, expressed as LC_{50}. Unless otherwise stated hardness and alkalinity expressed as mg $litre^{-1}$ $CaCO_3$. Lack of information in table reflects absence of these details in the original reference

Species	*Lifestage*	*Test conditions*	*Metal salt added*	*Response criteria*	*Metal concentration (mg $litre^{-1}$)*	*Reference*
NICKEL						
Acroneuria lycorias	Nymph	Static	$NiSO_4$	96-h LC_{50}	33·5	Warnick and Bell (1969)
Ephemerella subvaria	Nymph	Hardness 44, pH 7·25		96-h LC_{50}	4·0	
Daphnia magna	< 24-h old	Static, with food, hardness 45·3, pH 7·74	$NiCl_2$	48-h LC_{50}	1·12	Biesinger and Christensen (1972)
		Static, without food, hardness 45·3, pH 7·74		48-h LC_{50}	0·51	
Nais sp.		Static		24-h LC_{50}	16·2	Rehwoldt *et al.* (1973)
		Hardness 50, pH 7·6		96-h LC_{50}	14·1	
Gammarus sp.				24-h LC_{50}	15·2	
				96-h LC_{50}	13·0	
Trichopteran	Nymph			24-h LC_{50}	48·4	
				96-h LC_{50}	30·2	
Zygopteran	Nymph			24-h LC_{50}	26·4	
				96-h LC_{50}	21·2	
Chironomus sp.				24-h LC_{50}	10·2	
				96-h LC_{50}	8·6	
Amnicola sp.	Egg			24-h LC_{50}	26·0	
	Egg			96-h LC_{50}	11·4	
	Adult			24-h LC_{50}	21·2	
	Adult			96-h LC_{50}	14·3	
Cyclops abyssorum prealpinus	0·62 mm	Static	$NiCl_2$	48-h LC_{50}	15·0	Badouin and Scoppa (1974)
Eudiaptomus padanus padanus	0·42 mm	Hardness 0·6 meq $litre^{-1}$		48-h LC_{50}	3·6	
Daphnia hyalina	1·27 mm	pH 7·2		48-h LC_{50}	1·9	
Tubifex tubifex		Static with renewal Hardness 0·1, pH 6·3	$NiSO_4$	24-h LC_{50}	0·12	Brkovic-Popovic and Popovic (1977)

Table A7 — contd.

Species	*Lifestage*	*Test conditions*	*Metal salt added*	*Response criteria*	*Metal concentration (mg litre^{-1})*	*Reference*
				48-h LC_{50}	0·082	
		Hardness 34·2, pH 6·85		24-h LC_{50}	33·4	
				48-h LC_{50}	8·7	
		Hardness 34·2, pH 7·2		24-h LC_{50}	21·6	
				48-h LC_{50}	7·0	
		Hardness 261, pH 7·32		24-h LC_{50}	120·0	
				48-h LC_{50}	61·4	
COPPER						
Oreonectus rusticus	Adult	Continuous flow, fed	$CuSO_4$	96-h LC_{50}	3	Hubschman (1967*a*)
	Intermoult	Hardness 100–125,		24-h LC_{50}	6	
	Hatched young	pH 7·8–8·1		24-h LC_{50}	0·06–0·125	
Acroneuria lycorias	Nymph	Static	$CuSO_4$	96-h LC_{50}	8·3	Warnick and Bell (1969)
Ephemerella subvaria	Nymph	Hardness 44, pH 7·25		48-h LC_{50}	0·32	
Gammarus pseudolimnaeus		Continuous flow	$CuSO_4$	96-h LC_{50}	0·02	Arthur and Leonard (1970)
Campeloma decisum	11–27 mm			96-h LC_{50}	1·7	
Physa integra	4–7 mm			96-h LC_{50}	0·039	
Daphnia magna	< 24-h old	Static	$CuCl_2$			Biesinger and Christensen (1972)
		Hardness 45·3 fed		48-h LC_{50}	0·06	
		pH 7·4 not fed,		48-h LC_{50}	0·0098	
Nais sp.		Static		24-h LC_{50}	2·3	Rehwoldt *et al.* (1973)
		Hardness 50, pH 7·6		96-h LC_{50}	0·09	
Gammarus sp.				24-h LC_{50}	1·2	
				96-h LC_{50}	0·91	
Trichopteran	Nymph			24-h LC_{50}	12·1	
				96-h LC_{50}	6·2	
Zygopteran	Nymph			24-h LC_{50}	10·2	
				96-h LC_{50}	4·6	
Chironomous sp.				24-h LC_{50}	0·65	
				96-h LC_{50}	0·03	
Amnicola	Egg			24-h LC_{50}	4·5	
				96-h LC_{50}	9·3	

	Adult			24-h LC_{50}	1·5	
				96-h LC_{50}	0·9	
Cyclops abyssorum prealpinus	0·62 mm	Static	$CuCl_2$	48-h LC_{50}	2·5	Badouin and Scoppa (1974)
Eudiaptomus padanus padanus	0·42 mm	Hardness 0·6 meq litre^{-1},		48-h LC_{50}	0·5	
Daphnia hyalina	1·27 mm	pH 7·2		48-h LC_{50}	0·005	
Daphnia pulex	Multiaged	Static	$CuSO_4$	96-h LC_{50}	0·028	McIntosh and Keveren (1974)
Cyclops sp.	Multiaged	Hardness 112–157, pH 8·0–10·1		96-h LC_{50}	> 225	
Ephemerella grandis	Nymph	Continuous flow	$CuSo_4$	14-d LC_{50}	0·18–0·2	Nehring (1976)
Pteronarcys californica	Nymph	pH 6·3–7·2		14-d LC_{50}	10·1–13·9	
Tubifex tubifex		Static	$CuSO_4$	24-h LC_{50}	0·01	Brkovic-Popovic and Popovic (1977)
		Hardness 0·1, pH 6·3		48-h LC_{50}	0·0064	
		Hardness 34·2, pH 6·85		24-h LC_{50}	0·36	
				48-h LC_{50}	0·21	
		Hardness 34·2, pH 7·2		24-h LC_{50}	1·0	
				48-h LC_{50}	0·21	
		Hardness 261, pH 7·32		24-h LC_{50}	1·38	
				48-h LC_{50}	0·89	
Daphnia magna	< 24-h old	Static, fed trout pellets	$CuSO_4$	72-h LC_{50}	0·083	Winner *et al.* (1977)
		Alkalinity 2·0–2·36, fed algae, pH 7·2–9·5		72-h LC_{50}	0·085	
Gammarus fasciatus	Multiaged	Static, Hardness 206, pH 7·75	$CuSO_4$	48-h LC_{50}	0·21	Judy (1979)
Paramecium tetraurelia	Active growth phase	Static, pH 5·8–6·0 12°C	$CuSo_4$	24-h LC_{50}	0·0117	Szeto and Nyberg (1979)
		20°C		24-h LC_{50}	0·0104	
		27°C		24-h LC_{50}	0·0069	
		34°C		24-h LC_{50}	0·0047	
Viviparus bengalensis	2·5 cm	Static	$CuSO_4$			Gupta *et al.* (1981)
		Hardness 205, pH 7·6		24-h LC_{50}	2·15	
				48-h LC_{50}	1·27	
				96-h LC_{50}	0·066	
		Hardness 195, pH 7·9		24-h LC_{50}	0·54	
				48-h LC_{50}	0·093	
				96-h LC_{50}	0·06	

Table A7 — contd.

Species	*Lifestage*	*Test conditions*	*Metal salt added*	*Response criteria*	*Metal concentration (mg litre^{-1})*	*Reference*
		Hardness 180, pH 7·4		24-h LC_{50}	1·33	
				48-h LC_{50}	0·27	
				96-h LC_{50}	0·088	
		Hardness 190, pH 7·7		24-h LC_{50}	13·93	
				48-h LC_{50}	7·8	
				96-h LC_{50}	0·39	
Asellus meridianus		Static, River Hayle	$CuSO_4$	48-h LC_{50}	1·65, 1.7, 2·5	Brown (1977c)
		Hardness 25, R. Gannel		48-h LC_{50}	1·9	
		R. Bradwell		48-h LC_{50}	1·2	
ZINC						
Physa heterostropha		Static, 20°C, hardness 38–61, fed	$ZnCl_2$	LC_{50} 96-h	0·79–1·27	Cairns and Scheiers (1958)
Physa heterostropha		Static, 30°C, hardness 35–47, fed		LC_{50} 96-h	0·62–0·78	
Physa heterostropha		Static, 20°C, hardness 141–190, fed		LC_{50} 96-h	2·66–5·57	
Physa heterostropha		Static, 30°C, hardness 171–180, fed		LC_{50} 96-h	2·36–6·36	
Daphnia magna		20–22°C, pH 7·4–8·0, hardness 275–290	$ZnCl_2$	LC_{50} 48-h	9·8	Cabejszek and Stasiak (1960)
Daphnia magna				LC_{50} 96-h	7·7	
Daphnia magna			$ZnSO_4$	LC_{50} 48-h	10·8	
Daphnia magna				LC_{50} 96-h	7·4	
Asellus communis		Static, 21°C, hardness 20, pH 7·3	$ZnSO_4$	LC_{50} 96-h	8·7	Wurtz and Bridges (1961)
Asellus communis		Static, 21°C, hardness 100, pH 7·8		LC_{50} 96-h	12·7	
Agria sp.	Larvae	Static, 21°C, hardness 20, pH 7·3		LC_{50} 96-h	40·7	
Physa heterostropha	Adult	Static, 21°C, hardness 20, pH 7·3		LC_{50} 96-h	1·11	

Organism	Age/size	Test conditions	Compound	Criterion	Concentration	Reference
Physa heterostropha	Adult	Static, 21°C, hardness 100, pH 7·8		LC_{50} 96-h	3·16	
Limnodrilus hoffmeisteri		Static, 21°C, hardness 100, pH 7·8		LC_{50} 96-h	2·3	
Physa heterostropha	3–6 min	Static, 10·5°C, hardness 20, pH 7·3	$ZnSO_4$	LC_{50} 96-h	1·34	Wurtz (1962)
Physa heterostropha	3–6 min	Static, 32°C, hardness 20, pH 7·3		LC_{50} 96-h	1·55	
Physa heterostropha	3–6 min	Static, 10·5°C, hardness 100, pH 7·8		LC_{50} 96-h	1·92	
Physa heterostropha	3–6 min	Static, 32°C, hardness 100, pH 7·8		LC_{50} 96-h	4·9	
Helisonia campanulatum	Adult	Static, 13°C, hardness 20, pH 7·3		LC_{50} 96-h	3·85	
Helisonia campanulatum	Adult	Static 23°C, hardness 20, pH 7·3		LC_{50} 96-h	5·6	
Helisonia campanulatum	Adult	Static 13°C, hardness 100, pH 7·8		LC_{50} 96-h	13·4	
Helisonia campanulatum	Adult	Static, 23°C, hardness 100, pH 7·8		LC_{50} 96-h	5·6	
Daphnia magna		Hardness 180–214, 21°C, alkalinity 150–175, pH 7·1–7·4		LC_{50} 24-h	13·6	Malacea and Gruia (1965)
Daphnia magna				LC_{50} 48-h	2·5	
Daphnia magna				LC_{50} 75-h	1·0	
Physa heterostropha		Static, 20°C, hardness 43	$ZnCl_2$	LC_{50} 96-h	0·79–1·27	Patrick *et al.* (1968)
Acroneuria lycorias	Larvae	Static, hardness 44, alkalinity 40, pH 7·25	$ZnSO_4$	50% survival 14 d	32	Warnick and Bell (1969)
Ephemerella subvaria				50% survival 10 d	16	
Hydropsyche betteni				50% survival 11 d	32	
Daphnia magna	12 ± 12 h	pH 7·4–8·2, hardness 44–53 Alkalinity 41–50	$ZnCl_2$	LC_{50} 48-h	0·1	Biesinger and Christensen (1972)
Daphnia magna				LC_{50} 48-h	0·28	
Daphnia magna				LC_{50} 3-wk	0·16	
Nais sp.		Static, 17°C, hardness 50, pH 7·6	$ZnCl_2$	LC_{50} 24-h	21·2	Rehwoldt *et al.* (1973)

Table A7 — contd.

Species	Lifestage	Test conditions	Metal salt added	Response criteria	Metal concentration (mg litre^{-1})	Reference
Nais sp.				LC_{50} 96-h	18·4	Rehwoldt *et al.* (1973)—contd.
Gammarus sp.				LC_{50} 24-h	10·2	
Gammarus sp.				LC_{50} 96-h	8·1	
Trichopteran sp.	Larvae			LC_{50} 24-h	62·6	
Trichopteran sp.	Larvae			LC_{50} 96-h	58·1	
Zygopteran sp.	Larvae			LC_{50} 24-h	32	
Trichopteran sp.	Larvae			LC_{50} 96-h	26·2	
Chironomus sp.	Larvae			LC_{50} 24-h	21·5	
Chironomus sp.	Larvae			LC_{50} 96-h	18·2	
Amnicola sp.	Eggs			LC_{50} 24-h	28·1	
Amnicola sp.	Eggs			LC_{50} 96-h	20·2	
Amnicola sp.	Adults			LC_{50} 24-h	16·8	
Amnicola sp.	Adults			LC_{50} 96-h	14·0	
Daphnia hyalina	1·27 mm	Static, filtered lake water, 10°C, alkalinity 23, pH 7·2	$ZnSO_4$	LC_{50} 48-h	0·04	Badouin and Scoppa (1974)
Cyclops abyssorum prealpinus	0·62 mm	Static, filtered lake water, 10°C, alkalinity 23, pH 7·2		LC_{50} 48-h	5·5	
Eudiaptomus padarus padarus	0·43 mm	Static, filtered lake water, 10°C, alkalinity 23, pH 7·2		LC_{50} 48-h	0·5	
Paratya tasmaniensis	22·4 mm	Static, 15°C, hardness 109	$ZnSO_4$	LC_{50} 96-h	1·2	Thorp and Lake (1974)
Ephemeroptera grandis	Larvae	Flow through, 3–9°C, hardness 30–70 pH 7·0–7·2	$ZnSO_4$	LC_{50} 14-d	9·2	Nehring (1977)
Pteronarcys californica				LC_{50} 14-d	13·9	

Species	Stage	Conditions	Compound	Test	Value	Reference
Tubifex tubifex		Static, 20°C, hardness ~0·1, alkalinity, pH 6·3	$ZnSO_4$	LC_{50} 24-h	0·12	Brkovic-Popovic and Popovic (1977)
Tubifex tubifex				LC_{50} 48-h	0·11	
Tubifex tubifex		Static, 20°C, hardness 34·2, alkalinity 7·5, pH 6·8		LC_{50} 24-h	4·62	
Tubifex tubifex				LC_{50} 48-h	2·98	
Tubifex tubifex		Static, 20°C, hardness 34·2, alkalinity 22·5, pH 7·2		LC_{50} 24-h	3·64	
Tubifex tubifex				LC_{50} 48-h	2·57	
Tubifex tubifex		Static, 20°C, hardness 261, alkalinity 234, pH 7·3		LC_{50} 24-h	75·8	
Tubifex tubifex				LC_{50} 48-h	60·2	
Daphnia magna		Static, tap water, 5°C, hardness 45, pH 7·5	$ZnSO_4$	LC_{50} 48-h	2·3	Cairns *et al.* (1978)
Daphnia pulex				LC_{50} 48-h	1·6	
Nitocris sp.				LC_{50} 48-h	4·8	
Philodina acuticornis		Static, culture medium, 5°C, hardness 45, pH 7·5		LC_{50} 48-h	1·55	
Aeolosoma headlyeii				LC_{50} 48-h	18·1	
Daphnia magna		Static, tap water, 25°C, hardness 45, pH 7·5		LC_{50} 48-h	0·56	
Daphnia pulex				LC_{50} 48-h	0·28	
Nitrocis sp.				LC_{50} 48-h	1·65	
Philodina acuticornis		Static, culture medium, 25°C, hardness 45, pH 7·5		LC_{50} 48-h	0·5	
Aeolosoma headlyeii		Static, tap water, 25°C, hardness 45, pH 7·5		LC_{50} 48-h	13·5	
Daphnia magna		Static, hardness 116, alkalinity 76, pH 6·0, fed	$ZnSO_4$	LC_{50} 48-h	0·20	LeBlanc (1982)
Daphnia magna					0·24	
LEAD						
Daphnia pulex	Young	River water, pH 7·5, hardness 215, 23°C	$Pb(NO_3)_2$	LC_{50} 48-h	5·0	Bringmann and Kuhn (1959)
Daphnia magna		pH 7·4–8·0, 20–22°C, hardness 275–290	$PbCl_2$	LC_{50} 48-h	9·5	Cabejszek and Stasiak (1960)
Daphnia magna				LC_{50} 96-h	4·2	
Daphnia magna				LC_{50} 120-h	3·3	
Tubifex tubifex		Static, 20°C, pH 6·5	$Pb(NO_3)_2$	LC_{50} 24-h	49·0	Whitely (1968)
Limnodrilus hoffmeisteri						
Limnodrilus hoffmeisteri		Static, 20°C, pH 7·5		LC_{50} 24-h	> 50·0	
Limnodrilus hoffmeisteri		Static, 20°C, pH 8·5		LC_{50} 24-h	27·5	

Table A7 — contd.

Species	*Lifestage*	*Test conditions*	*Metal salt added*	*Response criteria*	*Metal concentration (mg litre^{-1})*	*Reference*
Acroneuria lycorias	Larvae	Static, pH 7·25, hardness 44, alkalinity 40	$PbSO_4$	50% survival > 14 d	64·0	Warnick and Bell (1969)
Ephemerella subvaria				50% survival > 7 d	16·0	
Hydropsyche betteni				50% survival > 7 d	32·0	
Daphnia magna	12 ± 12 h	Static, 18°C, hardness 45	$PbCl_2$	LC50 48-h	0·45	Biesinger and Christensen (1972)
Daphnia magna				LC_{50} 3-wk	0·3	
Daphnia hyalina	1·27 mm	Static, filtered lake water, 10°C, pH 7·2, alkalinity 23	$Pb(CH_3COO)_2$	LC_{50} 48-h	0·6	Badouin and Scoppa (1974)
Cyclops abyssoruum prealpinus	0·62 mm			LC_{50} 48-h	5·5	
Eudiaptomus padarus padarus	0·43 mm			LC_{50} 48-h	4·0	
Asellus meridianus	4–6 mm	Static, 20°C, hardness 25	$Pb(NO_3)_2$	LC_{50} 48-h	0·28–3·50	Brown (1977*c*)
Ephemeroptera grandis	Larvae	Flow through, 3–9°C, pH 7·0–7·2, hardness 30–70	$Pb(NO_3)_2$	LC_{50} 14-d	3·5	Nehring (1976)
Pteronarcys californica				LC_{50} 14-d	> 19·2	
Gammarus pseudolimnaeus	5–7 mm	Flow through, 15°C, pH 7·1–7·7, hardness 40–48 alkalinity 40–44	$Pb(NO_3)_2$	LC_{50} 28-d	0·0284	Spehar *et al.* (1978)
Gammarus pseudolimnaeus				LC_{50} 96-h	0·124	
Daphnia magna		Static, pH 6·0, hardness 116, alkalinity 7·6, fed	$Pb(NO_3)_2$	LC_{50} 48-h	0·12	LeBlanc (1982)
Daphnia magna					0·15	

Table A8

Figs 7.5–7.7. Effect of nickel, copper, zinc and lead on fish, expressed as LC_{50}. Unless otherwise stated hardness and alkalinity expressed as mg $litre^{-1}$ $CaCO_3$. Lack of information reflects absence of these details in the original reference

Species	*Lifestage*	*Test conditions*	*Metal salt added*	*Response criteria*	*Metal concentration (mg $litre^{-1}$)*	*Reference*
NICKEL						
Salmo gairdneri		Hardness 320, pH 7·62		48-h LC_{50}	76	Brown (1968)
Salmo gairdneri	1 yr	Static with renewal, hardness 240, pH 7·3–7·5	$NiSO_4$	48-h LC_{50}	32	Brown and Dalton (1970)
Fundulus diaphanus	20 cm	Static, hardness 53, pH 7·8	$Ni(NO_3)_2$	24-h LC_{50}	63·2	Rehwoldt *et al.* (1971)
				48-h LC_{50}	50·8	
				96-h LC_{50}	46·2	
Roccus saxatilis				24-h LC_{50}	10·0	
				48-h LC_{50}	8·4	
				96-h LC_{50}	6·2	
Lepomis gibbosus				24-h LC_{50}	16·4	
				48-h LC_{50}	12·0	
				96-h LC_{50}	8·1	
Roccus americanus				24-h LC_{50}	18·4	
				48-h LC_{50}	16·2	
				96-h LC_{50}	13·6	
Anguilla rostrata				24-h LC_{50}	14·0	
				48-h LC_{50}	13·2	
				96-h LC_{50}	13·0	
Cyprinus carpio				24-h LC_{50}	38·2	
				48-h LC_{50}	29·1	
				96-h LC_{50}	10·6	
Salmo gairdneri	2 months	Continuous flow, alkalinity 82–132, pH 6·4–8·3, field experiment	$Ni(NO_3)_2$	96-h LC_{50}	35·5	Hale (1977)
Cyprinus carpio	Hatching eggs	Static with renewal hardness 128, pH 7·4	$NiSO_4$	72-h LC_{50}	6·1	Blaylock and Frank (1979)
	1-day-old larvae			72-h LC_{50}	8·46	
	1-day-old larvae			96-h LC_{50}	6·16	

Table A8 — contd.

Species	Lifestage	Test conditions	Metal salt added	Response criteria	Metal concentration (mg litre^{-1})	Reference
COPPER						
Salmo gairdneri		Static, hardness 320, pH 7·62		48-h LC_{50}	0·4	Brown (1968)
Pimephales promelas	Adult	Static	$CuSO_4$	96-h LC_{50}	0·43	Mount (1968)
		Continuous flow		96-h LC_{50}	0·47	
Salmo gairdneri	1 yr	Static with renewal, hardness 240, pH 7·3–7·5	$CuSO_4$	48-h LC_{50}	0·75	Brown and Dalton (1970)
Salvelinus fontinalis	Juvenile (14-months)	Continuous flow, hardness 45·4, pH 7·5		96-h LC_{50}	0·09, 0·1	McKim *et al.* (1978)
Fundulus diaphanus	20 cm	Static hardness 53, pH 7·8	$Cu(NO_3)_2$	24-h LC_{50}	1·5	Rehwoldt *et al.* (1971)
				48-h LC_{50}	0·92	
				96-h LC_{50}	0·86	
Roccus saxatilis				24-h LC_{50}	8·3	
				48-h LC_{50}	6·2	
				96-h LC_{50}	4·3	
Lepomis gibbosus				24-h LC_{50}	3·8	
				48-h LC_{50}	2·9	
				96-h LC_{50}	2·4	
Roccus americanus				24-h LC_{50}	11·8	
				48-h LC_{50}	8·0	
				96-h LC_{50}	6·2	
Anguilla rostrata				24-h LC_{50}	10·6	
				48-h LC_{50}	8·2	
				96-h LC_{50}	6·4	
Cyprinus carpio				24-h LC_{50}	2·1	
				48-h LC_{50}	1·0	
				96-h LC_{50}	0·81	
Salmo salar	Parr	Static with renewal, hardness 8–10, pH 6·5–6·7		96-h LC_{50}	0·125	Wilson (1972)
Salmo gairdneri	32 mm, 0·36 g	Alkalinity 100, pH 6·5	$CuSO_4$	24-h LC_{50}	0·14	Shaw and Brown (1974)
				48-h LC_{50}	0·12	
		Alkalinity 100, pH 7·5		24-h LC_{50}	0·13	
				48-h LC_{50}	0·11	

Bluntnose minnow	Juvenile	Static, hardness 316		48-h LC_{50}	19	Andrew (1976)
		hardness 255		48-h LC_{50}	9·6	
		hardness 205		48-h LC_{50}	9·1	
		hardness 158		48-h LC_{50}	3·3	
Pimephales promelas	Adult	Static, hardness 120–310, pH 8·0–8·5	$CuSO_4$	96-h LC_{50}	0·6–0·98	Brungs *et al.* (1976)
Lebistes reticulatus	Juvenile	Continuous flow, hardness 66·8–98·0, pH 7·0				Chynoweth *et al.* (1976)
		Control		96-h LC_{50}	0·112, 0·138	
		+ NTA		96-h LC_{50}	0·224	
		+ Humic acid		96-h LC_{50}	0·188	
		+ EDTA		96-h LC_{50}	0·184	
		+ Glycine		96-h LC_{50}	0·183	
		+ Sewage effluent		96-h LC_{50}	0·099	
Noemacheilus barbatulus	Adult	Continuous flow, hardness 249, pH 8·26	$CuSO_4$	63-d LC_{50}	0·25	Solbé and Cooper (1976)
Oncorhynchus tshawytscha	Newly hatched swim up alevins	Continuous flow, hardness 23, pH 7·1		96-h LC_{50}	0·026	Chapman (1978)
				96-h LC_{50}	0·019	
	5–8 month parr smolt			96-h LC_{50}	0·038	
				96-h LC_{50}	0·026	
Salmo gairdneri	Newly hatched swim up alevins			96-h LC_{50}	0·028	
				96-h LC_{50}	0·017	
	5–8 months parr smolt			96-h LC_{50}	0·018	
				96-h LC_{50}	0·029	
Salmo gairdneri		Continuous flow	$CuSO_4$			Howarth and Sprague (1978)
	2·2 g	hardness 31, pH 6·0		96-h LC_{50}	0·0289	
	2·7 g	hardness 30, pH 9·0		96-h LC_{50}	0·03	
	3·2 g	hardness 101, pH 6·0		96-h LC_{50}	0·04	
	1·0 g	hardness 98, pH 9·0		96-h LC_{50}	0·0859	
	1·7 g	hardness 371, pH 6·0		96-h LC_{50}	0·0822	
	3·1 g	hardness 364, pH 9·0		96-h LC_{50}	0·1	

Table A8 — contd.

Species	Lifestage	Test conditions	Metal salt added	Response criteria	Metal concentration (mg litre^{-1})	Reference
ZINC						
Salmo gairdneri	3–15 months	Static, 17·5°C, hardness 12, pH 6·6–6·7	$ZnSO_4$	LC_{50} 48-h	0·6	Lloyd (1960)
Salmo gairdneri		Static, 17·5°C, hardness 50, pH 7·0–7·2		LC_{50} 48-h	2	
Salmo gairdneri		Static, 17·5°C, hardness 320, pH 7·6–7·8		LC_{50} 48-h	4	
Salmo gairdneri	15–17 cm long	Flow through, bore-hole water, hardness 504, pH 7·18, 15·5°C	$ZnSO_4$	LC_{50} 48-h	4·76	Solbé (1974)
Noemocheilus barbatulus	6·5–11 cm long	Bore-hole water, static, hardness 297, pH 7·44, 13·9°C		LC_{50} 48-h	4	Solbé and Flook (1975)
Noemocheilus barbatulus	6·5–11 cm long	Bore-hole water, static, hardness 291, pH 7·66, 11·8°C		LC_{50} 48-h	6·5	
Salmo salar	Parr	Flow through, pH 6·1–6·7, 10°C, hardness 12·1–24·4, alkalinity < 1–11	$ZnSO_4$	LC_{50} 21-d		Farmer *et al.* (1979)
Salmo salar	4·8 g				1·45	
Salmo salar	5·5 g				1·60	
Salmo salar	9·2 g				0·51	
Salmo salar	9·6 g				1·46	
Salmo salar	10·1 g				0·34	
Salmo salar	16·7 g				0·35	
Salmo gairdneri	19·2 cm long	pH 7·3–7·9, 14·5–15·8°C	$ZnSO_4$	LC_{50} 5-d	4·6	Ball (1967)
Abramis brama	10·6 cm long	pH 7·2–7·9, 10·0–12·6°C			14·3	
Perca fluviatilis	13·6 cm long	pH 7·2–7·8, 10·4–13·8°C			16·0	
Rutilus rutilus	9·0 cm long	pH 7·2–7·9, 10·0–12·6°C			17·3	
Rutilus rutilus	8·6 cm long	pH 7·2–7·8, 10·4–13·2°C			17·3	

LEAD						
Lebistes reticulatus		pH 7·8–8·2, 25–27°C, hardness 80, alkalinity 51	$Pb(NO_3)_2$	LC_{50} 30-d	5	Crandal and Goodnight (1962)
Salmo gairdneri	86 mm long	Static, 14°C, pH 8·15, hardness 385, alkalinity 267	$Pb(NO_3)_2$	LC_{50} 96-h	1·32 (Dissolved)	Davies *et al.* (1976)
Salmo gairdneri				LC_{50} 96-h	542 (Total)	
Salmo gairdneri	130 mm long	Static, 7°C, pH 8·78, hardness 290, alkalinity 228		LC_{50} 96-h	1·47 (Dissolved)	
Salmo gairdneri						
Salmo gairdneri	161 mm long	Flow through, 7°C, pH 6·85, hardness 30, alkalinity 29		LC_{50} 14-d	0·20 (Total)	
Salmo gairdneri	145 mm long	Flow through, 10°C, pH 6·85 hardness 32, alkalinity 30		LC_{50} 96-h	1·17 (Total)	

Table A9

Fig. 8.4. Number of species of algae recorded at naturally highly-acid habitats

Site	*pH*	*Number of species*	*Reference*
Lake Katanuma (Japan)	1·8	3	Satake and Saijo (1974)
Smoking Hills (Canada)			
Pond 1	1·8	6	Sheath *et al.* (1982)
Pond 2	2·8	8	
Pond 3	3·6	10	
Pond 4	2·0	3	
Kootenay Paint Pots (Canada)			
Site 1	3·2	10	Wehr and Whitton (1983*a*)
Site 2	4·7	9	
Site 3	3·2	10	
Site 4	4·0	9	
Site 5	3·1	10	
Site 6	3·4	10	

Table A10

Fig. 8.5. Lowest pH at which different phyla of freshwater algae are found. Species names as recorded in original references

Species	*pH of site at which found*	*Comments*	*Location*	*Reference*
Chlorophyceae				
Chlamydomonas aplanata var *acidophila*	1·8–3·0 (av. 2·7)	Found in 25% of 52 sites with pH ⩽ 3, more common in late summer[f]	England	Hargreaves *et al.* (1975)
Chlamydomonas acidophila	Minimum 1·8		[a]	Whitton and Diaz (1981)
Chlamydomonas globosa	3·0		USA	Bennett (1969)[e]
Chlamydomonas sp.	1·8–7·2	11 out of 43 occurrences ⩽ pH 3, very decided tolerance of highly acid conditions	USA[b]	Lackey (1938)
Chlamydomonas sp.	2·7–2·9 (av. 2·8)	Found at 11% of 52 sites with ⩽ 3	England	Hargreaves *et al.* (1975)
Chlamydomonas sp.	⩽ 3·0		USA	Steinback (1966); Bennett (1969)
Chlamydomonas sp.	Minimum 2·8 median[c]		W. Va, USA	Warner (1971)
Chlamydomonas sp.	Minimum 2·25		[a]	Whitton and Diaz (1981)
Ulothrix zonata	2·4–7·0	13 out of 19 ⩽ pH 3, common in some streams, very decided tolerance for highly acid conditions	USA[b]	Lackey (1938)
Ulothrix zonata	3·0	Not found in late summer survey[f], only one occurrence	England	Hargreaves *et al.* (1975)

Table A10 — contd.

Species	*pH of site at which found*	*Comments*	*Location*	*Reference*
Ulothrix variabilis	≤ 4·1		Arizona, USA	Lampkin and Sommerfield (1982)
Ulothrix subtilis	≤ 3·0		USA	Bennett (1969)
Ulothrix tenerrima	≤ 3·0		USA	Steinback (1966); Bennett (1969)
Ulothrix tenerrima	2·6–3·0		Pa, USA	Koryak *et al.* (1972)
Ulothrix tenerrima	Minimum 2·8 median	Found at all 15 sites sampled, including site with ferric deposit	W. Va, USA	Warner (1971)
Stigeoclonium subtile	3·8 median	Only found at one acid site[d]	W. Va, USA	Warner (1971)
Stigoeclonium sp.	av. < 4·5		Pa, USA	Weed and Rutschky (1972)
Stigoeclonium sp.	4·0–6·4	Only 1 occurrence < pH 5	USA[b]	Lackey (1938)
Mougeotia sp.	2·4–6·2	2 out of 6 occurrences ≤ pH 3, occasional distribution	USA[b]	Lackey (1938)
Mougeotia sp.	≤ 3·0		USA	Bennett (1969)
Desmidium sp.	1·8–3·2	1 out of 5 occurrences ≤ pH 3, sparing distribution	USA[b]	Lackey (1938)
Zygogonium ericetorum	2·5–3·0 (av. 2·8)	Found at 19% of 52 sites with pH ≤ 3, easily recognizable macroscopically sometimes forming areas of continuous cover	England	Hargreaves *et al.* (1975)
Zygogonium ericetorum	Minimum 3·5 median		W. Va, USA	Warner (1971)
Zygogonium sp.	Minimum 2·10		Widespread[a]	Whitton and Diaz (1981)

Cladophora sp.	Minimum 4·5 median	Only found at one acid site [d]	W. Va, USA	Warner (1971)
Microthamnion strictissimum	2·9	Not found in winter survey, only one occurrence [f]	England	Hargreaves *et al.* (1975)
Microthamnion strictissimum	Minimum 2·9 median	Found at 13 of 15 sites sampled, including site with ferric deposit	W. Va, USA	Warner (1971)
Microthamnion kuetzingianum	Minimum 3·25		[a]	Whitton and Diaz (1981)
Microthamnion kuetzingianum	⩽ 4·1		Arizona, USA	Lampkin and Sommerfield (1982)
Chlorogonium elongatum	⩽ 3·0		USA	Bennett (1969)
Penium jenneri	⩽ 3·0		USA	Bennett (1969)
Penium margaritaceum	Minimum 4·9 median	Only founnd at one acid site [d]	W. Va, USA	Warner (1971)
Microspora pachyderma	Minimum 3·3 median	Found at 12 of 15 sites sampled, including site with ferric deposit	W. Va, USA	Warner (1971)
Closterium acerosum	Minimum 3·3 median	Found at 10 of 15 sites sampled	W. Va, USA	Warner (1971)
Cosmarium spp.	Minimum 3·3 median	Including site with ferric deposit	W. Va, USA	Warner (1971)
Staurastrum alternans	Minimum 3·3 median		W. Va, USA	Warner (1971)
Characium sp.	1·5–3·0 (av. 2·6)	Found at 13% of 52 sites with pH ⩽ 3	England	Hargreaves *et al.* (1975)
Scenedesmus abundans	Minimum 3·3 median		W. Va, USA	Warner (1971)
Spirogyra sp.	Minimum 3·3 median		W. Va, USA	Warner (1971)
Protococcus viridis	Minimum 3·3 median		W. Va, USA	Warner (1971)
Cylindrocapsa geminella var. *minor*	Minimum 3·6 median		W. Va, USA	Warner (1971)
Ankistrodesmus falcatus	Minimum 3·4 median		W. Va, USA	Warner (1971)
Closteriopsis longissima	Minimum 3·3 median		W. Va, USA	Warner (1971)
Binucleria tetrana	Minimum 3·6 median		W. Va, USA	Warner (1971)
Oocystis sp.	Minimum 4·5 median		W. Va, USA	Warner (1971)

Table A10 — contd.

Species	*pH of site at which found*	*Comments*	*Location*	*Reference*
Pediastrum sp.	Minimum 4·9 median	Only found at one acid site[d]	W. Va, USA	Warner (1971)
Gloeocystis gigas	Minimum 4·9 median	Only found at one acid site[d]	W. Va, USA	Warner (1971)
Chlorella sp.	Minimum 0·9		[a]	Whitton and Diaz (1981)
Hormidium rivulare	2·5–3·0 (2·8 av.)	Found at 23% of 52 sites with pH $\leqslant 3$, recognizable macroscopically	England	Hargreaves *et al.* (1975)
Hormidium rivulare	Minimum 2·30	Widespread distribution	[a]	Whitton and Diaz (1981)
Hormidium pseudostichococcus	Minimum 2·25		[a]	Whitton and Diaz (1981)
Hormidium vulcanum	Minimum 2·60	In thermal streams	USA	Whitton and Diaz (1981)
Stichococcus bacillaris	1·8–2·8	Found at 6% of 52 sites with pH $\leqslant 3$,	England	Hargreaves *et al.* (1975)
Stichococcus sp.	Minimum 1·80	Cells rounded, $>1 \leqslant 2\,\mu m$	[a]	Whitton and Diaz (1981)
Stichococcus sp.	Minimum 1·41	Cells rounded, $>2\,\mu m$	[a]	Whitton and Diaz (1981)
Stichococcus sp.	Minimum 2·50	Cells ± cylindrical, $\leqslant 2\,\mu m$	[a]	Whitton and Diaz (1981)
Stichococcus sp.	Minimum 0·90	Cells ± cylindrical, $>2\,\mu m \leqslant 4\,\mu m$	[a]	Whitton and Diaz (1981)
Desmids	4·0	Rare distribution in a strip mine lake	Ohio, USA	Riley (1960)
Chrysophyceae				
Chromulina sp.	1·8–6·6	6 out of 12 occurrences $\leqslant$ pH 3, very decided tolerance for highly acid conditions	USA[b]	Lackey (1938)

Chromulina sp.	≤ 3·0		USA	Bennett (1969)
Dinobryon sp.	3·2–7·2	Only one occurrence < pH 5	USA[b]	Lackey (1938)
Phaetothamnion sp.	2·6–3·8	3 out of 7 occurrences ≤ pH 3	USA[b]	Lackey (1938)
Gloeochrysis turfosa	1·8–3·0 (2·7 av.)	Found at 61% of 52 sites with pH < 3, easily recognizable macroscopically sometimes forming areas of continuous cover. More common in late summer,[f] motile and non-motile forms, non-motile more common	England	Hargreaves *et al.* (1975)
Ochromonas sp.	Minimum 3·5 median		W. Va, USA	Warner (1971)
Chrysophyte	Minimum 3·70	Palmelloid > 8 μm long	[a]	Whitton and Diaz (1981)
Bacillariophyceae (Diatoms)				
Navicula nivalis	3·0	Not found in winter survey, only one occurrence[f]	England	Hargreaves *et al.* (1975)
Navicula mutica	Minimum 2·52		[a]	Whitton and Diaz (1981)
Navicula spp.	1·8–7·2	13 out of 42 occurrences ≤ pH 3, abundant, very decided tolerance for highly acid conditions	USA[b]	Lackey (1938)
Navicula sp.	2·5 + 2·9	Found at only one of 52 sites with pH ≤ 3	England	Hargreaves *et al.* (1975)

Table A10 — contd.

Species	*pH of site at which found*	*Comments*	*Location*	*Reference*
Navicula sp.	≤ 3·0		USA	Steinback (1966)
Navicula sp.	Minimum 3·3 median	Including site with ferric deposit	W. Va, USA	Warner (1971)
Navicula sp.	Minimum 2·40		[a]	Whitton and Diaz (1981)
Navicula sp.	4·0	Abundant distribution in a strip mine lake	Ohio, USA	Riley (1960)
Tabellaria sp.	3·7–7·2	Only 2 of 4 occurrences < pH 5	USA[b]	Lackey (1938)
Pinnularia acoricola	1·5–3·0 (2·8 av.)	Found at 71% of 52 sites with pH ≤ 3	England	Hargreaves *et al.* (1975)
Pinnularia microstauron	2·5 + 2·9	Found at only one of 52 sites with pH ≤ 3	England	Hargreaves *et al.* (1975)
Pinnularia termitina	Minimum 2·8 median	Found at all 15 sites sampled including site with ferric deposit	W. Va, USA	Warner (1971)
Pinnularia termitina	Minimum 3·80		[a]	Whitton and Diaz (1981)
Pinnularia biceps	< 3·5 av.	Dominant in strip lake	Ohio, USA	Fritz and Carlson (1982)
Pinnularia hilseana var. *hilseana*	Minimum 3·3 median		W. Va, USA	Warner (1971)
Pinnularia subcapitata	Minimum 3·30		[a]	Whitton and Diaz (1981)
Pinnularia braunii	Minimum 3·10		[a]	Whitton and Diaz (1981)
Pinnularia variabilis	Minimum 3·70		[a]	Whitton and Diaz (1981)
Pinnularia sp.	Minimum 3·05	Not *Pinnularia acoricola*	[a]	Whitton and Diaz (1981)

Nitzschia subcapitellata	2·5–3·0 (2·7 av.)	Found at 32% of 52 sites with pH $\leqslant 3$	England	Hargreaves *et al.* (1975)
Nitzschia elliptica var. *alexandrina*	2·5–3·0 (2·8 av.)	Found at 31% of 52 sites with pH $\leqslant 3$, more common in winter.[f] Morphologically different forms noted	England	Hargreaves *et al.* (1975)
Nitzschia palea	2·7 + 2·9	Found at only one of 52 sites with pH $\leqslant 3$	England	Hargreaves *et al.* (1975)
Nitzschia palea	Minimum 2·50		[a]	Whitton and Diaz (1981)
Nitzschia hungarica	Minimum 3·5 median	Including site with ferric deposit	W. Va, USA	Warner (1971)
Nitzschia ovalis	2·5–2·8 (2·6 av.)	Found at only 2 of 52 sites with pH $\leqslant 3$	England	Hargreaves *et al.* (1975)
Nitzschia pusilla	Minimum 2·50	Widespread distribution	[a]	Whitton and Diaz (1981)
Nitzschia thermalis	Minimum 3·10		[a]	Whitton and Diaz (1981)
Nitzschia sp. type A	2·5–3·0 (2·8 av.)	Found at 3 and 1 site(s) respectively from 52 possible sites with pH $\leqslant 3$, both are possibly just variants of known spp., but difficult to identify (applies to type A and type B)	England	Hargreaves *et al.* (1975)
Nitzschia sp. type B	2·5			
Eunotia exigua	2·5–3·0 (2·7 av.)	Found at 27% of 52 sites with pH $\leqslant 3$, morphologically different forms noted	England	Hargreaves *et al.* (1975)
Eunotia exigua	$\leqslant 3$		USA	Steinback (1966); Warner (1968); Bennett (1969)

Table A10 — contd.

Species	*pH of site at which found*	*Comments*	*Location*	*Reference*
Eunotia exigua	Minimum 2·8 median	Found at all 15 sites sampled, including site with ferric deposit	W. Va, USA	Warner (1971)
Eunotia exigua	Minimum 2·20	Widespread distribution	[a]	Whitton and Diaz (1981)
Eunotia sudetica	Minimum 4·5 median	Only found at one acid site[d]	W. Va, USA	Warner (1971)
Eunotia lunaris	Minimum 3·10		[a]	Whitton and Diaz (1981)
Eunotia tenella	≤ 4·1	Abundant distribution, found adjacent to mine seep	Arizona, USA	Lampkin and Sommerfield (1982)
Eunotia sp.	4·0	Rare distribution in a strip mine lake	Ohio, USA	Riley (1960)
Frustulia rhomboides	≤ 3		USA	Bennett (1969)
Frustulia rhomboides	≤ 3·9	Abundant distribution in an impoundment	Alabama, USA	Dills and Rogers (1974)
Frustulia rhomboides	Minimum 3·1	Not var. *saxonica*	[a]	Whitton and Diaz (1981)
Frustulia rhomboides var. *saxonica*	Minimum 2·8 median	Found at 14 of 15 sites sampled, including site with ferric deposit	W. Va, USA	Warner (1971)
Synedra rumpens	≤ 3		USA	Bennett (1969)
Synedra ulna	< 4·1	Infrequent occurrence	Arizona, USA	Lampkin and Sommerfield (1982)
Synedra acus	Minimum 3·6 median		W. Va, USA	Warner (1971)
Achnanthes minutissima	≤ 3·9	Abundant distribution in an impoundment	Alabama, USA	Dills and Rogers (1974)

Achnanthes marginulata	Minimum 3·3 median		W. Va, USA	Warner (1971)
Achnanthes lanceolata var. *dubia*	≤ 4·1	Infrequent occurrence	Arizona, USA	Lampkin and Sommerfield (1982)
Diatoma hiemale	Minimum 4·9 median	Only found at one acid site[d]	W. Va, USA	Warner (1971)
Diatoma sp.	4·0	Rare distribution in a strip mine lake	Ohio, USA	Riley (1960)
Surirella ovata	Minimum 3·3 median	Found at 10 of 15 sample sites including site with ferric deposit	W. Va, USA	Warner (1971)
Comphonema olivaceum	Minimum 3·10		[a]	Whitton and Diaz (1981)
Comphonema parvulum	Minimum 3·10		[a]	Whitton and Diaz (1981)
Comphonema sp.	Minimum 2·8 median	Found at 10 of 15 sample sites including site with ferric deposit	W. Va, USA	Warner (1971)
Fragilaria sp.	Minimum 2·8 median		W. Va, USA	Warner (1971)
Cyclotella glomerata	Minimum 2·8 median		W. Va, USA	Warner (1971)
Tabellaria floculosa	Minimum 3·6 median		W. Va, USA	Warner (1971)
Meridion circulare	Minimum 3·3 median		W. Va, USA	Warner (1971)
Stephanodiscus hantzschii	Minimum 3·3 median		W. Va, USA	Warner (1971)
Melosira granulata	Minimum 3·3 median	Only found at one acid site[d]	W. Va, USA	Warner (1917)
Melosira distans	Minimum 3·10		[a]	Whitton and Diaz (1981)
Cocconeis diminuta	Minimum 3·8 median	Only found at one acid site[d]	W. Va, USA	Warner (1971)
Rhoicosphenia curvata	Minimum 3·5 median	Only found at one acid site,[d]	W. Va, USA	Warner (1971)
Stauroneis anceps	Minimum 3·5 median	Only found at one acid site[d]	W. Va, USA	Warner (1971)
Epithemia sorex	≤ 4·1	Infrequent occurrence	Arizona, USA	Lampkin and Sommerfield (1982)
Neidium alpinum	Minimum 3·70		[a]	Whitton and Diaz (1981)

Table A10 — contd.

Species	*pH of site at which found*	*Comments*	*Location*	*Reference*
Anomoeneis exilis	Minimum 3·10		[a]	Whitton and Diaz (1981)
Diatoms	Acid conditions	Generally easily recognizable macroscopically, sometimes forming areas of continuous cover		Hargreaves *et al.* (1975)
Cryptophyceae				
Cryptomonas erosa	1·8–7·2	10 out of 21 occurrences below pH 5, only one occurrence below pH 3	USA[b]	Lackey (1938)
Cryptomonas ovata	⩽ 3·0		USA	Bennett (1969)
Cryptomonas sp.	2·6–2·9	Found at two out of 52 sites with pH ⩽ 3, not *erosa* or *ovata*	England	Hargreaves *et al.* (1976)
Cryptomonas sp.	Minimum 2·50		[a]	Whitton and Diaz (1981)
Xanthophyceae				
Tribonema affine	⩽ 4·1		Arizona, USA	Lampkin and Sommerfield (1982)
Euglenophyceae				
Euglena mutabilis	1·8–7·2	34 out of 65 occurrences ⩽ pH 3, very abundant, very decided tolerance for highly acid conditions, cells apparently devoid of a flagellum	USA[b]	Lackey (1938)

Euglena mutabilis	1·5–3·0 (2·6 av.)	Found at 90% of 52 sites with pH $\leqslant 3$, easily recognizable macroscopically sometimes forming areas of continuous cover, often most abundant	England	Hargreaves *et al.* (1975)
Euglena mutabilis	$\leqslant$ 3·0		USA	Steinback (1966); Warner (1968); Bennett (1969)
Euglena mutabilis	Minimum 2·8 median	Found at all 15 sites sampled including site with ferric deposit	W. Va, USA	Warner (1971)
Euglena mutabilis	1·41	Most widespread spp. noted	[a]	Whitton and Diaz (1981)
Euglena sp.	4·9–7·2	Only one occurrence below pH 5	USA[b]	Lackey (1938)
Euglena sp.	$\leqslant$ 3·0		USA	Bennett (1969)
Euglena sp.	4·0	Coal strip mine lake, rare distribution	Ohio, USA	Riley (1960)
Lepocinclis ovum	3·0–7·2	Only 4 out 7 occurrences below pH 5	USA[b]	Lackey (1938)
Lepocinclis ovum	Minimum 2·5		[g]	Whitton and Diaz (1981)
Trachelomonas sp.	Minimum 2·9 median		W. Va, USA	Warner (1971)
Phacus sp.	Minimum 4·9 median	Only found at one acid site[d]	W. Va, USA	Warner (1971)
Cyanobacteria (Myxophyceae)				
Oscillatoria sp.	Minimum 2·8 median	Including site with ferric deposit	W. Va, USA	Warner (1971)
Anabaena sp.	Minimum 4·5 median		W. Va, USA	Warner (1971)
Chroococcus rufescens	Minimum 4·5 median		W. Va, USA	Warner (1971)
Porphyriosiphon notarissii	Minimum 3·6 median	Only found at one acid site[d]	W. Va, USA	Warner (1971)

Table A10 — contd.

Species	*pH of site at which found*	*Comments*	*Location*	*Reference*
Amphitrix janthina	Minimum 4·9 median	Only found at one acid site[d]	W. Va, USA	Warner (1971)
Lyngbya sp.	Minimum 4·9 median	Only found at one acid site[d]	W. Va, USA	Warner (1971)
Spirulina nordstedtii	Minimum 4·9 median	Only found at one acid site[d]	W. Va, USA	Warner (1971)
Myxophyte	4·0	Unknown genus, rare distribution in a strip mine lake	Ohio, USA	Riley (1960)

[a] From a survey of sites in Belgium, Eire, Great Britain and the USA — most of the sites were associated with mine drainage.
[b] From a survey of 92 sites in Indiana and West Virginia, USA.
[c] pH values in Warner's (1971) study are median pH values for each site. Between 22 and 167 analyses were carried out for each site.
[d] Only 13 of Warner's 15 sites had median pH values below 5·0 and these are taken as acid sites.
[e] Bennett (1969) surveyed eight 'acid creeks', six of which were below pH 3·0.
[f] Hargreaves *et al.* (1975) carried out two surveys of the flora: the first in August to early October ('late summer') and the second in the latter part of February ('winter').
[g] Parsons (1968) noted communities in streams which received intermittent acid mine drainage. His pH readings were taken at different times of the year to give an indication of the range of conditions found at each site. Only two communities ('A' and 'B') had pH values below 5·0 (on average) throughout the year; only these two have been used.

Table A11

Fig. 8.8. Lowest pH at which different taxa of freshwater invertebrates are found. Species names as recorded in original references. See notes after Table A10

Species	*pH of site at which found*	*Comments*	*Location*	*Reference*
Ephemeroptera				
Ephemerella sp.	Minimum 4·5 median		W. Va, USA	Warner (1971)
Hexagenia sp.	Minimum 4·5 median		W. Va, USA	Warner (1971)
Stenonema sp.	Minimum 4·5 median		W. Va, USA	Warner (1971)
Ameletus sp.	Minimum 4·9 median	Only found at one continuously acid site[d]	W. Va, USA	Warner (1971)
Blasturus sp.	Minimum 4·9 median	Only found at one acid site[d]	W. Va, USA	Warner (1971)
Leptophlebia sp.	Minimum 4·9 median	Only found at one acid site[d]	W. Va, USA	Warner (1971)
Paraleptophlebia sp.	Minimum 4·5 median	Only found at one acid site[d]	W. Va, USA	Warner (1971)
Ephemeroptera	Minimum 2·3	A few larvae	USA[b]	Lackey (1938)
Ephemeroptera	3·3 av.	Ferric deposits	Pa, USA	Koryak *et al.* (1972)
Plecoptera				
Acroneuria sp.	Minimum 4·9 median	Only found at one acid site[d]	W. Va, USA	Warner (1971)
Allocapnia sp.	Minimum 4·9 median	Only found at one acid site[d]	W. Va, USA	Warner (1971)
Alloperla sp.	Minimum 4·5 median	Only found at one acid site[d]	W. Va, USA	Warner (1971)
Brachyptera sp.	Minimum 4·5 median		W. Va, USA	Warner (1971)
Isoperla sp.	Minimum 4·5 median	Only found at one acid site[d]	W. Va, USA	Warner (1971)
Leuctra sp.	Minimum 4·5 median		W. Va, USA	Warner (1971)
Nemoura sp.	Minimum 4·5 median		W. Va, USA	Warner (1971)
Plecoptera	3·8 av		Pa, USA	Koryak *et al.* (1972)
Odonata				
Pantala hymenea	2·8–4·6	Bicarbonate level = 0, temperature = 25–29°C	Missouri, USA[g]	Parsons (1968)
Odonata	Minimum 3·8 median		W. Va, USA	Warner (1971)
Hemiptera				
Cerridae	< 4·0	In area of coal slurry ponds	Illinois, USA	Vinikour (1979)

Table A11 — contd.

Species	*pH of site at which found*	*Comments*	*Location*	*Reference*
Cerridae	Minimum 4·5 median		W. Va, USA	Warner (1971)
Corixidae	Minimum 3·3 median	Only found at one acid site[d]	W. Va, USA	Warner (1971)
Coleoptera				
Esolus parallelepipedus	3·9	Low density, marked iron precipitate, Zn: 0·26 mg litre^{-1}	England	Armitage (1979)
Lanternarius brunneus	< 4·0	Not a truly aquatic species	Illinois, USA	Vinikour (1979)
Chelonarium sp.	4·3	Fe: 1.5 mg litre^{-1} SO_4: 156 mg litre^{-1}	Pa, USA	Weed and Rutschky (1972)
Psephenus sp.	4·3	Fe: 1.5 mg litre^{71}, SO_4: 156 mg litre^{-1} — immature forms	Pa, USA	Weed and Rutschky (1972)
Berosus sp.	4·3	Fe: 1·5 mg litre^{-1}, SO_4: 156 mg litre^{-1} — immature forms	Pa, USA	Weed and Rutschky (1972)
Laccophilus sp.	4·3	Fe: 1·5 mg litre^{-1}, SO_4: 156 mg litre^{-1} — immature forms	Pa, USA	Weed and Rutschky (1972)
Gyrinidae	< 4·9	Near surface of lakes	USA[b]	Lackey (1938)
Dytiscidae	Minimum 2·8 median	Found at 9 of 15 sites sampled, including site with ferric precipitate	W. Va, USA	Warner (1971)
Coleoptera	3·3 av.	Ferric deposits	Pa, USA	Koryak *et al.* (1972)
Hegaloptera				
Sialis sp.	3·2 av.		Pa, USA	Tomkiewicz and Dunson (1977)
Sialis sp.	Minimum 2·8 median	Found at 13 of 15 sites sampled, including site with ferric precipitate	W. Va, USA	Warner (1971)
Sialis sp.	4·3	Fe: 1·5 mg litre^{-1}, SO_4: 156 mg litre^{-1}	Pa, USA	Weed and Rutschky (1972)
Sialis sp.	2·8–4·6	Bicarbonate level = 0, temperature = 25–29°C	Missouri, USA	Parsons (1968)
Chauloides sp.	Minimum 2·8 median		W. Va, USA	Warner (1971)
Chauloides sp.	4·2	Bicarbonate level = 0, temperature = 28°C	Missouri, USA	Parsons (1968)

Nigronia sp.	4·3	Fe: 1·5 mg litre^{-1}, SO_4: 156 mg litre^{-1}	Pa, USA	Weed and Rutschky (1972)
Corydalus sp.	4·3	Fe: 1·5 mg litre^{-1}, SO_4: 156 mg litre^{-1}	Pa, USA	Weed and Rutschky (1972)
Trichoptera				
Ptilostomis sp.	3·2 av.		Pa, USA	Tomkiewicz and Dunson (1977)
Ptilostomis sp.	Minimum 2·8 median		W. Va, USA	Warner (1971)
Lype sp.	4·3	Fe: 1·5 mg litre^{-1}, SO_4: 156 mg litre^{-1}	Pa, USA	
Cheumatopsyche sp.	4·2	Bicarbonate level = 0, temperature = 28°C	Missouri, USA	Parsons (1968)
Cheumatopsych sp.	Minimum 4·9 median	Only found at one acid site[d]	W. Va, USA	Warner (1971)
Hydropsyche sp.	Minimum 4·9 median	Only found at one acid site[d]	W. Va, USA	Warner (1971)
Polycentropus sp.	Minimum 4·5 median	Only found at one acid site[e]	W. Va, USA	Warner (1971)
Psycomyia sp.	Minimum 4·9	Only found at one acid site[d]	W. Va, USA	Warner (1971)
Pycnopsyche sp.	Minimum 3·8 median		W. Va, USA	Warner (1971)
Rhyacophila sp.	Minimum 4·9 median	Only found at one acid site[d]	W. Va, USA	Warner (1971)
Trichoptera (larvae)	Minimum 2·3	In a stream	USA[b]	Lackey (1938)
Lepidoptera				
Synclita sp.	Minimum 3·3 median	Only found at one acid site[d]	W. Va, USA	Warner (1971)
Diptera				
Corethra sp.	< 4·9	Under stones	USA[b]	Lackey (1938)
Chironomus sp.	< 4·9	In strip mine (flooded) pits	USA[b]	Lackey (1938)
Chironomus sp.	Minimum 2·8 median	Found at 12 of 15 sites sampled, including site with ferric precipitate	W. Va, USA	Warner (1971)
Rhypholophus sp.	3·9	Low density, marked iron precipitate, Zn: 0·26 mg litre^{-1}	England	Armitage (1979)
Eriocera sp.	Minimum 3·5 median	Only found at one acid site,[d] site with ferric precipitate	W. Va, USA	Warner (1971)
Pedicia sp.	Minimum 3·5 median	Including site with ferric precipitate	W. Va, USA	Warner (1971)
Tipula sp.	Minimum 3·5 median	Including site with ferric precipitate	W. Va, USA	Warner (1971)
Antocha sp.	3·3 av.	Ferric deposits	Pa, USA	Koryak *et al.* (1972)
Antocha sp.	Minimum 4·5 median	Only found at one acid site,[d]	W. Va, USA	Warner (1971)
Hexatoma sp.	Minimum 4·5 median	Only found at one acid site[d]	W. Va, USA	Warner (1971)
Pseudolimnophila sp.	Minimum 4·9 median	Only found at one acid site[d]	W. Va, USA	Warner (1971)

Table A11 — contd.

Species	*pH of site at which found*	*Comments*	*Location*	*Reference*
Pseudolimnophila sp.	4·2	Bicarbonate level = 0, temperature = 28°C	Missouri, USA	Parsons (1968)
Tendipes gr. riparius	2·6–3·0	Abundant	Pa, USA	Koryak *et al.* (1972)
Tendipes attenuatus	4·3 av.	Large numbers in July and August. Fe: 1·5 mg litre^{-1}	Pa, USA	Weed and Rutschky (1972)
Simulium sp.	Minimum 4·5 median		W. Va, USA	Warner (1971)
Atherix sp.	Minimum 4·5 median	Only found at one acid site[d]	W. Va, USA	Warner (1971)
Psychoda sp.	Minimum 3·3 median	Only found at one acid site[d]	W. Va, USA	Warner (1971)
Psychoda sp.	3·3 av.	Ferric deposits	Pa, USA	Koryak *et al.* (1972)
Procladius sp.	2·8–4·6	Bicarbonate level = 0, temperature = 25–29°C	Missouri, USA	Parsons (1968)
Probezzia sp.	2·8–4·6	Bicarbonate level = 0, temperature = 25–29°C	Missouri, USA	Parsons (1968)
Tabanus sp.	4·2	Bicarbonate level = 0, temperature = 28°C	Missouri, USA	Parsons (1968)
Palpomyia sp.	4·3 av.	Fe: 3·0 mg litre^{-1}	Pa, USA	Weed and Rutschky (1972)
Culicidae	Minimum 2·4	Mosquito larvae, in a quiet pool	USA[b]	Lackey (1938)
Culicidae	Minimum 4·9 median	Only found at one acid site[d]	W. Va, USA	Warner (1971)
Orthocladiinae	3·9	Marked iron precipitate, low density distribution, Zn: 0·26 mg litre^{-1}	England	Armitage (1979)
Sarcophagidae	3·3 av.	Ferric deposits	Pa, USA	Koryak *et al.* (1972)
Chironomidae	3·2 av.		Pa, USA	Tomkiewicz and Dunson (1977)
Chironomidae	Minimum 2·8 median	One genus, not *Chironomus* found at 12 of 15 sample sites	W. Va, USA	Warner (1971)
Ceratopogonidae	Minimum 3·3 median		W. Va, USA	Warner (1971)
Tendipedids	3·3 av.	Ferric deposits	Pa, USA	Koryak *et al.* (1972)

References

Aanes, K.J. (1980). A preliminary report from a study on the environmental impact of pyrite mining and dressing in a mountain stream in Norway. pp. 419–42. In: Flannagan, J.F. & Marshall, K.E. (Eds) *Advances in Ephemeroptera Biology*. Plenum Press, New York.

Abdullah, M.I. & Royle, L.G. (1972). Heavy metal content of some rivers and lakes in Wales. *Nature (London)*, **238**, 329–30.

Abdullah, M.I., Banks, J.W., Miles, D.L. & O'Grady, K.T. (1976). Environmental dependence of manganese and zinc in the scales of Atlantic Salmon, *Salmo salar* (L.) and Brown trout *Salmo trutta* (L.). *Freshwater Biology*, **6**, 161–6.

Abel, P.D. & Green, D.W.J. (1981). Ecological and toxicological studies on invertebrate fauna of two rivers in the Northern Pennine orefield. pp. 109–22. In: Say, P.J. & Whitton, B.A. (Eds) *Heavy Metals in Northern England: Environmental and Biological Aspects*. Department of Botany, University of Durham, England.

Abo-Rady, M.D.K. (1979). Gehalt an Schwermetallen (Cd, Cu, Hg, Ni, Pb, Zn) in Bachforellen im Leine-Raum Göttingen. *Zeitschrift für Lebensmittel Untersuchung und Forschung*, **168**, 259–63.

Adams, T.G., Atchison, G.J. & Vetter, R.J. (1980). The impact of an industrially contaminated lake on heavy metal levels in its effluent stream. *Hydrobiologia*, **69**, 187–93.

Allen, M.B. (1959). Studies with *Cyanidium caldarium*, an anomalously pigmented chlorophyte. *Archiv für Microbiologie*, **32**, 270–7.

American Public Health Association (1981). *Standard Methods for the Examination of Water and Wastewater*. 15th Edition. 1134 pp. American Public Health Association Inc., New York.

Anderson, R.V. (1977). Concentration of cadmium, copper, lead and zinc in thirty-five genera of freshwater macroinvertebrates from the Fox River, Illinois and Wisconsin. *Bulletin of Environmental Contamination and Toxicology*, **18**, 345–9.

Andrew, R.W. (1976). Toxicity relationships to copper forms in natural waters. pp. 127–43. In: Andrew, R.W., Hodson, P.V. & Konasewich, D.E. (Eds)

Toxicity to Biota of Metal Forms in Natural Waters. Proceedings of Workshop, Duluth, Minnesota, October 1975.

Apel, W.A., Dugan, P.R., Filppi, J.A. & Rheins, M.S. (1976). Detection of *Thiobacillus ferrooxidans* in acid mine environments by indirect fluorescent antibody staining. *Applied and Environmental Microbiology*, **32**, 159–65.

Apel, W.A., Dugan, P.R. & Tuttle, J.H. (1980). Adenosine 5′-triphosphate formation in *Thiobacillus ferrooxidans* vesicles by H^+ ion gradients comparable to those of environmental conditions. *Journal of Bacteriology*, **142**, 295–301.

Armitage, P.D. (1979). The effects of mine drainage and organic enrichment on benthos in the River Nent system, Northern Pennines. *Hydrobiologia*, **74**, 119–28.

Armitage, P.D. & Blackburn, J.H. (1985). Chironomidae in a Pennine stream system receiving mine drainage and organic enrichment. *Hydrobiologia*, **121**, 165–72.

Arndt, U. (1974). The Kautsky-effect: a method for the investigation of the actions of air pollutants in chloroplasts. *Environmental Pollution*, **6**, 181–94.

Arora, A. & Gupta, A.B. (1983). The effect of copper sulphate on formation of separation discs in *Oscillatoria* sp. *Archiv für Hydrobiologie*, **96**, 261–6.

Arthur, J.W. & Leonard, E.N. (1970). Effects of copper on *Gammarus pseudolimnaeus*, *Physa integra* and *Campeloma decisum* in soft water. *Journal of the Fisheries Research Board of Canada*, **27**, 1277–83.

Ashuckian, S.H. & Finlayson, B.J. (1979). Safe zinc and copper levels from the Spring Creek drainage for steel head trout and in the Upper Sacramento River, California. *California Fish and Game*, **65**, 80–99.

Astruc, M. (1979). Metal forms and speciation. pp. 439–45. In: *Proceedings of the International Conference on Management and Control of Heavy Metals in the Environment*. CEP Consultants, Edinburgh.

Atchison, G.J., Henry, M.G. & Sandheinrich, M.B. (1987). Effects of metals on fish behavior: a review. *Environmental Biology of Fish*, **18**, 11–25.

Aulio, K. & Salin, M. (1982). Enrichment of copper, zinc, manganese, and iron in five species of pondweeds (*Potamogeton* spp.). *Bulletin of Environmental Contamination and Toxicology*, **29**, 320–5.

Austin, A. & Munteanu, N. (1984). Evaluation of changes in a large oligotrophic wilderness park lake exposed to mine tailing effluent for 14 years: the phytoplankton. *Environmental Pollution Series A*, **33**, 39–62.

Babich, H., Shopsis, C. & Borenfreund, E. (1986). *In vitro* cytotoxicity testing of aquatic pollutants (cadmium, copper, zinc, nickel) using established fish cell lines. *Ecotoxicology and Environmental Safety*, **11**, 91–9.

Baccini, P., Hohl, H. & Bundi, T. (1978). Phenomenology and modelling of heavy metal distribution in lakes. *Verhandlungen, Internationale Vereinigung für Theoretische und Angewandte Limnologie*, **20**, 1971–5.

Bache, J.J. (1987). *World Gold Deposits. A Geological Classification*. 178 pp. Translated by J. Wanklyn. North Oxford Academic, London.

Badouin, M.F. & Scoppa, P. (1974). Acute toxicity of various metals to freshwater zooplankton. *Bulletin of Environmental Contamination and Toxicology*, **12**, 745–51.

Badsha, K.S. & Goldspink, C.R. (1982). Preliminary observations on the heavy metal content of four species of freshwater fish in north-west England. *Journal of Fish Biology*, **21**, 251–67.

Bagchi, S.N., Karamchandani, A. & Bisen, P.S. (1985). Isolation and preliminary characterization of cadmium and lead tolerant strains of the cyanobacterium *Lyngbya* species IW 487. *Microbios Letters,* **29**, 65–8.

Ball, I.R. (1967). The relative susceptibilities of some species of freshwater fish to poisons. II. Zinc. *Water Research,* **1**, 777–83.

Barnes, H.L. & Romberger, S.B. (1968). The chemical aspects of acid mine drainage. *Journal of the Water Pollution Control Federation,* **40**, 371–84.

Bartlett, L., Rabe, F.W. & Funk, W.H. (1974). Effects of copper, zinc and cadmium on *Selanastrum capricornutum. Water Research,* **8**, 179–85.

Barton, P. (1978). The acid mine drainage. pp. 313–58. In: Nriagu, J.O. (Ed.) *Sulphur in the Environment. II. Ecological Aspects.* Wiley-Interscience, New York.

Battarbee, R.W. (1984). Diatom analysis and the acidification of lakes. *Philosophical Transactions of the Royal Society of London Series B,* **305**, 451–77.

Beardall, J. (1981). CO_2 accumulation by *Chlorella saccharophila* (Chlorophyceae) at low external pH: evidence for active transport of inorganic carbon at the chloroplast envelope. *Journal of Phycology,* **17**, 371–3.

Bell, H.L. (1971). Effect of low pH on the survival and emergence of aquatic insects. *Water Research,* **5**, 313–19.

Bendell-Young, L.I., Harvey, H.H. & Young, J.F. (1986). Accumulation of cadmium by White Suckers (*Catostromus commersoni*) in relation to fish growth and lake acidification. *Canadian Journal of Fisheries and Aquatic Science,* **43**, 806–11.

Bennett, H.D. (1969). Algae in relation to mine water. *Castanea,* **34**, 306–28.

Besch, K.W. & Roberts-Pichette, P. (1970). Effects of mining pollution on vascular plants in the Northwest Miramichi River System. *Canadian Journal of Botany,* **48**, 1647–56.

Besch, K.W., Richard, M. & Cantin, R. (1972). Benthic diatoms as indicators of mining pollution in the Northwest Miramichi River System, New Brunswick, Canada. *Internationale Revue der Gesamten Hydrobiologie,* **57**, 39–74.

Biesinger, K.E. & Christensen, G.M. (1972). Effects of various metals on survival, growth, reproduction and metabolism of *Daphnia magna. Journal of the Fisheries Research Board of Canada,* **29**, 1691–1700.

Bird, S.C. (1987). The effect of hydrological factors on trace metal contamination in the River Tawe, South Wales. *Environmental Pollution,* **45**, 87–124.

Birtwell, I.K., Hartman, G.F., Anderson, B., McLeay, D.J. & Malick, J.G. (1984). *A brief investigation of Arctic grayling* (Thymallus arcticus) *and aquatic invertebrates in the Minto Creek drainage, Mayo, Yukon Territory: an area subjected to placer mining.* xi + 57 pp. Canadian Technical Report on Fisheries and Aquatic Sciences No. 1287.

Bjerklie, D.M. & LaPerriere, J.D. (1985). Gold mining effects on stream hydrology and water quality, Circle Quadrangle, Alaska. *Water Resources Bulletin,* **21**, 235–43.

Blaylock, B.G. & Frank, M.L. (1979). A comparison of the toxicity of nickel to the developing eggs and larvae of carp. *Bulletin of Environmental Contamination and Toxicology,* **21**, 604–11.

Boavida, M.J. & Heath, R.T. (1986). Phosphatase activity of *Chlamydomonas acidophila* Negoro (Volvocales, Chlorophyceae). *Phycologia,* **25**, 400–4.

Bonaly, J., Miginiac-Maslow, M. Brochiero, E., Hoarau, A. & Mestre, J.C. (1986). Cadmium effects on the energetics of *Euglena* during the development of cadmium resistance. *Journal of Plant Physiology*, **123**, 349–58.

Borgman, U. (1980). Interactive effects of metals in mixtures on biomass production kinetics of freshwater copepods. *Canadian Journal of Fisheries and Aquatic Science*, **37**, 1295–1302.

Bosman, D.J. (1983). Lime treatment of acid mine water and associated solids/liquid separation. pp. 71–84. In: Odendaal, P.E. (Ed.) *Mine Water Pollution Proceeds — International Association of Water Pollution Research*. Water Science and Technology (South Africa) **15**(2), 1–181.

Bowen, H.J.M. (1966). *Trace Elements in Biochemistry*. 241 pp. Academic Press, London.

Bowen, H.J.M. (1975). Residence time of heavy metals in the environment. pp. 1–19. In: Hutchinson, T.C. (Ed.) *Symposium Proceedings of the International Conference on Heavy Metals in the Environment Volume 1: Plenary, Futures, Analytical*. Toronto, Ontario.

Bowen, H.J.M. (1985). The cycles of copper, silver and gold. pp. 1–27. In: Hutzinger, O. (Ed.) *The Handbook of Environmental Chemistry Volume I Part D. The Natural Environment and the Biogeochemical Cycles*. Springer-Verlag, Berlin.

Bowen, R. & Gunatilaka, A. (1977). *Copper, its Geology and Economics*. Applied Science Publishers, London.

Boyle, E.A. (1979). Copper in Natural Waters. pp. 77–88. In: Nriagu, J.O. (Ed.) *Copper in the Environment Part 1: Ecological Cycling*. Wiley-Interscience, New York.

Bradley, S.B. & Lewin, J. (1982). Transport of heavy metals on suspended sediments under high flow conditions in a mineralized region of Wales. *Environmental Pollution Series B*, **4**, 257–67.

Bradshaw, A.D. & McNeilly, T. (1981). *Evolution and Pollution*. 76 pp. Institute of Biology Studies in Biology No. 130. Edward Arnold, London.

Bringmann, G. & Kuhn, R. (1959). Vergleichende wasser-toxikologische Untersuchungen an Bakterian, Algen und Kleinkrebsen. *Gesundheitsingenieur*, **80**, 115–20.

British Geological Survey (1985). *World Mineral Statistics 1979–1983*. 275 pp. Her Majesty's Stationary Office, London.

Brkovic-Popovic, I. & Popovic, M. (1977). Effects of heavy metals on survival and respiration rate of tubificid worms. Part 1: Effects on survival. *Environmental Pollution*, **13**, 65–71.

Brock, T.D. (1969). Microbial growth under extreme conditions. *Symposium of the Society of General Microbiology*, **19**, 15–41.

Brooks, R.R. & Rumsby, M.G. (1965). The biogeochemistry of trace element uptake by some New Zealand bivalves. *Limnology and Oceanography*, **10**, 521–7.

Brown, B.E. (1977*a*). Effects of mine drainage on the River Hayle, Cornwall. A) Factors affecting concentrations of copper, zinc and iron in water, sediments and dominant invertebrate fauna. *Hydrobiologia*, **52**, 221–33.

Brown, B.E. (1977*b*) Uptake of copper and lead by a metal-tolerant isopod *Asellus meridianus* Rac. *Freshwater Biology*, **7**, 235–44.

Brown, B.E. (1977*c*). Observations on the tolerance of the isopod *Asellus meridianus* Rac. to copper and lead. *Water Research*, **10**, 555–9.

Brown, B.E. (1978). Lead detoxification by a copper-tolerant isopod. *Nature (London)*, **276**, 388–90.

Brown, B.T. & Rattigan, B.M. (1979). Toxicity of soluble copper and other ions to *Elodea canadensis*. *Environmental Pollution*, **20**, 303–14.

Brown, D.H. & Beckett, R.P. (1985). Intracellular and extracellular uptake of cadmium by the moss *Rhytidiadelphus squarrosus*. *Annals of Botany, New Series*, **55**, 179–88.

Brown, V.M. (1968). The calculation of the acute toxicity of mixtures of poisons to rainbow trout. *Water Research*, **2**, 723–33.

Brown, V.M. & Dalton, R.A. (1970). The acute lethal toxicity to rainbow trout of mixtures of poisons. *Journal of Fish Biology*, **2**, 211–16.

Brugam, R.B. & Lusk, M. (1986). Diatom evidence for neutralization in acid surface mine lakes. pp. 115–29. In: Smol, J.P., Battarbee, R.W., Davies, R.B. & Meriläinen, J. (Eds) *Diatoms and Lake Acidity*. Junk, Dordrecht, Boston & Lancaster.

Brungs, W.A., Geckler, J.R. & Gast, M. (1976). Acute and chronic toxicity of copper to the fathead minnow in a surface water of variable quality. *Water Research*, **10**, 37–43.

Burrows, I.G. & Whitton, B.A. (1983). Heavy metals in water, sediments and invertebrates from a metal-contaminated river free of organic pollution. *Hydrobiologia*, **106**, 263–73.

Burton, D.T., Jones, A.H. & Cairns, J. (1972). Acute zinc toxicity to rainbow trout (*Salmo gairdneri*): confirmation of the hypothesis that death is related to tissue hypoxia. *Journal of the Fisheries Research Board of Canada*, **29**, 1463–6.

Burton, M.A.S. & Peterson, P.J. (1979). Studies on Zn localization in aquatic bryophytes. *Bryologist*, **82**, 594–8.

Butler, P.A., Andren, L., Bonde, G.J., Jernelov, A. & Reisch, D.J. (1971). Monitoring Organisms. pp. 101–12, In: *Technical Conference on Marine Pollution and its Effects on Living Resources and Fishing, Rome 1970. Supplementary Methods of Detection, Measurement and Monitoring of Pollutants in the Marine Environment*.

Cabejszek, I. & Stasiak, M. (1960). Studies on the effects of metals on water biocenosis using the *Daphnia magna* index, part II. *Rocznik Paristwowego Zakladu Hygieny*, **11**, 533–40.

Cabrera, F., Toca, C.G., Diaz, E. & Arambarri, P. de (1984). Acid mine water and agricultural pollution in a river skirting the Doñana National Park (Guadiamar River, South West Spain). *Water Research*, **18**, 1469–82.

Caines, L.A., Watt, A.W. & Wells, D.E. (1985). The uptake and release of some trace metals by aquatic bryophytes in acidified waters in Scotland. *Environmental Pollution Series B*, **10**, 1–18.

Cairns, J. & Scheiers, A. (1958). The effects of temperature and hardness of water upon the toxicity of zinc to the pond snail *Physa heterostropha* (Say). *Notulae Naturae*, **308**, 1–11.

Cairns, J., Buikema, A.L., Heath, A.G. & Parker, B.C. (1978). Effects of temperature on aquatic organism sensitivity to selected chemicals. *Bulletin of the Virginia Water Resources Research Center No. 106*.

Cammarota, V.A. (1980). Production and uses of zinc. pp. 1–38. In: Nriagu, J.O. (Ed) *Zinc in the Environment Part 1: Ecological Cycling*. Wiley-Interscience, New York.

Campbell, P.G.C. & Stokes, P.M. (1985). Acidification and toxicity of metals to aquatic biota. *Canadian Journal of Fisheries and Aquatic Science*, **42**, 2034–49.

Campbell, P.G.C., Tessier, A. & Bisson, M. (1979). Anthropogenic influences on the speciation and fluvial transport of trace metals. pp. 453–6. In: *Proceedings of the International Conference on Management and Control of Heavy Metals in the Environment*. CEP Consultants, Edinburgh.

Carlson-Gunnoe, N.E., Law, C.B. & Bissonnette, G.K. (1983). *In situ* persistence of indicator bacteria in streams polluted with acid mine drainage. *Water Research*, **17**, 1119–24.

Carpenter, K.E. (1924). A study of the fauna of rivers polluted by lead mining in the Aberystwyth District of Cardiganshire. *Annals of Applied Biology*, **11**, 1–23.

Carpenter, K.E. (1926). The lead mine as an active agent in river pollution. *Annals of Applied Biology*, **13**, 395–401.

Carrithers, R.B. & Bulow, F.J. (1973). An ecological survey of the West Fork of the Obey River, Tennessee with emphasis on the effects of acid mine drainage. *Journal of the Tennessee Academy of Science*, **48**, 65–72.

Carson, R. (1962). *Silent Spring*. 304 pp. Hamish Hamilton, London.

Carter, J.R. (1972). Some observations on the diatom *Pinnularia acoricola* Hustedt. *Microscopy (London)*, **32**, 162–5.

Cassin, P.E. (1974). Isolation, growth and physiology of acidophilic chlamydomonads. *Journal of Phycology*, **10**, 439–47.

Chamier, A.-C. (1987). Effect of pH on microbial degradation of leaf litter in seven streams of the English Lake District. *Oecologia (Berlin)*, **71**, 491–500.

Chapman, G.A. (1978). Toxicities of cadmium, copper and zinc to four juvenile stages of Chinook salmon and steelhead. *Transactions of the American Fisheries Society*, **107**, 841–7.

Christensen, E.R., Scherfig, J. & Dixon, P.S. (1979). Effects of manganese, copper and lead on *Selanastrum capricornutum* and *Chlorella stigmatophora*. *Water Research*, **13**, 79–82.

Chynoweth, D.P., Black, J.A. & Maney, K.H. (1976). Copper toxicity to phytoplankton, as affected by organic ligands, other cations and inherent tolerance of algae to copper. pp. 145–7. In: Andrew, R.W. Hodson, P.V. & Konasewich, D.E. (Eds). *Toxicity to Biota of Metal Forms in Natural Waters*. Proceedings of Workshop, Duluth, Minnesota, October 1975.

Clark, C.J. & Crawshaw, D.H. (1979). A study into the treatability of ochreous mine-water discharges. *Water Pollution Control*, **78**, 446–62.

Clarke, S.E., Stuart, J. & Sanders-Loehr, J. (1987). Induction of siderophore activity in *Anabaena* spp. and its moderation of copper toxicity. *Applied and Environmental Microbiology*, **53**, 917–22.

Clymo, R.S. (1967). Control of cation concentrations, and in particular of pH, in *Sphagnum* dominated communities. pp. 273–84. In: Golterman, H.L. & Clymo, R.S. (Eds) *Chemical Environment in the Aquatic Habitat*. North Holland, Amsterdam.

Cochran, R.C. (1987). Effects of coal leachates on fish spermatogenesis. *Canadian Journal of Fisheries and Aquatic Science*, **44**, 134–9.

Collvin, L. (1985*a*). Effects of copper on growth and starvation in perch, *Perch fluviatilis* L. *Journal of Fish Biology*, **27**, 757–64.
Collvin, L. (1985*b*). The effect of copper on growth, food consumption and food conversion of perch *Perca fluviatilis* L. offered maximal food rations. *Aquatic Toxicology*, **6**, 105–13.
Correa, M. (1987). Physiological effects of metal toxicity on the tropical freshwater shrimp *Macrobrachium carcinus* (Linnea, 1758). *Environmental Pollution*, **45**, 149–55.
Cotton, F.A. & Wilkinson, G. (1972). *Advanced Inorganic Chemistry: A Comprehensive Text*. 1145 pp. Wiley, New York.
Cowx, I.G. (1982). Concentrations of heavy metals in the tissues of trout *Salmo trutta* and char *Salvelinus alpinus* from two lakes in North Wales. *Environmental Pollution Series A*, **29**, 101–10.
Crandal, C.A. & Goodnight, C.J. (1962). Effects of sublethal concentrations of several toxicants on growth of the common guppy *Lebistes reticulatus*. *Limnology and Oceanography*, **7**, 223–39.
Craze, B. (1977). Restoration of Captains Flat Mining Area. *Soil Conservation Journal of New South Wales*, **33**, 99–105.
Cushing, C.E. (1979). Trace elements in a Columbia River food web. *Northwest Science*, **53**, 118–25.
Cusimano, R.F., Brakke, D.F. & Chapman, G.A. (1986). Effects of pH on the toxicities of cadmium, copper, and zinc to Steelhead Trout (*Salmo gairdneri*). *Canadian Journal of Fisheries and Aquatic Science*, **43**, 1497–503.
Dach, H. von (1943). The effect of pH on pure cultures of *Euglena mutabilis*. *Ohio Journal of Science*, **43**, 47–8.
Dahl, J. (1963). Transformation of iron and sulphur compounds in soil, and its relation to Danish inland fisheries. *Transactions of the American Fisheries Society*, **92**, 260–4.
Dallinger, R., Prosi, F., Segner, H. & Back, H. (1987). Contaminated food and uptake of heavy metals by fish: a review and a proposal for future research. *Oecologia (Berlin)*, **73**, 91–8.
Davies, P.H., Goettl, J.P., Sinley, J.R. & Smith, N.F. (1976). Acute and chronic toxicity of lead to Rainbow Trout *Salmo gairdneri* in hard and soft water. *Water Research*, **10**, 199–206.
Davison, W., Hilton, J., Lishman, J.P. & Pennington, W. (1985). Contemporary lake transport processes determined from sedimentary records of copper mining activity. *Environmental Science and Technology*, **19**, 356–60.
Davison, W., De Mora, S.J., Harrison, R.M. & Wilson, S. (1987). pH and ionic strength dependence of the ASV response of cadmium, lead and zinc in solutions which simulate natural waters. *Science of the Total Environment*, **60**, 35–44.
DeCosta, J. & Preston, C. (1980). The phytoplankton productivity of an acidic lake. *Hydrobiologia*, **70**, 39–49.
Deniseger, J., Austin, A. & Lucey, W.P. (1986). Periphyton communities in a pristine mountain stream above and below heavy metal mining operations. *Freshwater Biology*, **16**, 209–18.
Denny, P. (1972). Sites of nutrient absorption in aquatic macrophytes. *Journal of Ecology*, **60**, 819–29.

Denny, P. (1981). Limnological studies on the relocation of lead in Ullswater, Cumbria. pp. 93–8. In: Say, P.J. & Whitton, B.A. (Eds) *Heavy Metals in Northern England: Environmental and Biological Aspects*. Department of Botany, University of Durham, England.

Denny, P. & Welsh, R.P.H. (1979). Lead accumulation in plankton blooms from Ullswater, the English Lake District. *Environmental Pollution*, **18**, 1–9.

Denys, L. (1984). *Achnanthes andicola* (Cl.) Hust. and *Pinnularia acoricola* Hust. (Bacillariophyceae) recorded in Belgium. *Bulletin de la Société Royale de Botanique de Belgique*, **117**, 73–9.

Dietz, F. (1973). The enrichment of heavy metals in submerged plants. pp. 53–60. In: Jenkins, S.H. *Advances in Water Pollution Research*. Sixth International Conference held in Jerusalem, June 8-23, 1972. Pergamon Press, Oxford & New York.

Dills, G. & Rogers, D.T. (1974). Macroinvertebrate community structure as an indicator of acid mine pollution. *Environmental Pollution*, **6**, 239–62.

Down, C.G. & Stocks, J. (1977). *Environmental Impact of Mining*. 371 pp. Applied Science Publishers, London.

Dugan, P.R. (1975). Bacterial ecology of strip mine areas and its relationship to the production of acidic mine drainage. *Ohio Journal of Science*, **75**, 266–79.

Dugan, P.R. (1984). Desulfurization of coal by mixed microbial cultures. pp. 3–9. In: Strohl, W.R. & Tuovinen, O.H. (Eds) *Microbial Chemoautotrophy*. Ohio State University Press, Columbus.

Dugan, P.R. (1986). Microbiological desulfurization of coal and its increased monetary value. *Biotechnology and Bioengineering Symposium*, **16**, 185–203.

Dugan, P.R. (1987*a*). Prevention of formation of acid drainage from high-sulphur coal refuse by inhibition of iron- and sulfur-oxidizing microorganisms. I. Preliminary experiments in controlled shaken flasks. *Biotechnology and Bioengineering*, **29**, 41–8.

Dugan, P.R. (1987*b*). Prevention of formation of acid drainage from high-sulfur coal refuse by inhibition of iron-and sulfur-oxidizing microorganisms. II. Inhibition in 'run of mine' refuse under simulated field conditions. *Biotechnology and Bioengineering*, **29**, 49–54.

Dugan, P.R., MacMillan, C.B. & Pfister, R.M. (1970). Aerobic heterotrophic bacteria indigenous to pH 2·8 acid mine water: microscopic examination of acid streamers. *Journal of Bacteriology*, **101**, 973–81.

Duke, J.M. (1980). Nickel in rocks and ores. In: Nriagu, J.O. (Ed.) *Nickel in the Environment*. Wiley-Interscience, New York.

Du Plessis, H.M. (1983). Using lime treated acid mine water for irrigation. pp. 145–54. In: Odendaal, P.E. (Ed.) *Mine Water Pollution Proceeds — International Association of Water Pollution Research*. Water Science and Technology (South Africa) **15**(2), 1–181.

Durum, W.H. & Haffty, J. (1963). Implications of the minor element content of some major streams in the world. *Geochimica et Cosmochimica Acta*, **27**, 1–11.

Eddlemon, G.K. & Tolbert, V.R. (1983). Chatanooga shale exploitation and the aquatic environment: the critical issues. *Environment International*, **9**, 85–95.

Edgington, D.N. & Robbins, J.A. (1976). Records of lead deposition in Lake Michigan sediments since 1800. *Environmental Science and Technology*, **10**, 266–74.

Ehrle, E.B. (1960). *Eleocharis acicularis* in acid mine drainage. *Rhodora* **62**, 95–7.

Elder, J.F. & Horne, A.J. (1978). Copper cycles and $CuSO_4$ algicidal capacity in two Californian lakes. *Environmental Management*, **2**, 17–30.

Elderfield, H., Thornton, I. & Webb, J.S. (1971). Heavy metals and oyster culture in Wales. *Marine Pollution Bulletin*, **2**, 44–7.

Elkington J. (1987). *The Green Capitalists. Industry's Search for Environmental Excellence*. 258 pp. Gollancz, London.

Ellwood, J.W., Hildebrand, S.G. & Beauchamp, J.J. (1976). Contributions of gut contents to the concentration and body burden of elements in *Tipula* spp. from a spring-fed stream. *Journal of the Fisheries Research Board of Canada*, **33**, 1930–8.

Emmons, W.H. (1918). *The Principles of Economic Geology*. 606 pp. McGraw-Hill, New York.

Enk, M.D. & Mathis, B.J. (1977). Distribution of cadmium and lead in a stream ecosystem. *Hydrobiologia*, **52**, 153–8.

Ernst, W.H.O. & Marquenie-van der Werff, M. (1978). Aquatic angiosperms as indicators of copper contamination. *Archiv für Hydrobiologie*, **83**, 356–66.

Everard, M. & Denny, P. (1984). The transfer of lead by freshwater snails in Ullswater, Cumbria. *Environmental Pollution Series A*, **35**, 299–314.

Everard, M. & Denny, P. (1985*a*). Flux of lead in submerged plants and its relevance to a freshwater system. *Aquatic Botany*, **21**, 181–93.

Everard, M. & Denny, P. (1985*b*). Particulates and the cycling of lead in Ullswater, Cumbria. *Freshwater Biology*, **15**, 215–26.

Eyres, J.P. & Pugh-Thomas, M. (1978). Heavy metal pollution of the River Irwell (Lancashire, U.K.) demonstrated by analysis of substrate materials and macroinvertebrate tissue. *Environmental Pollution*, **16**, 129–36.

Fängström, I. (1972). The effects of some chelating agents and their copper complexes on photosynthesis in *Scenedesmus quadricauda*. *Physiologia Plantarum*, **27**, 389–97.

Farmer, C.J., Ashfield, D. & Samant, H.S. (1979). Effects of zinc on juvenile atlantic salmon *Salmo salar*: acute toxicity, food intake, growth and bioaccumulation. *Environmental Pollution*, **19**, 103–17.

Fayed, S.E. & Abd-El-Shafy, H.I. (1985). Accumulation of Cu, Zn, Cd and Pb by aquatic macrophytes. *Environment International*, **11**, 77–87.

Felts, P.A. & Heath, A.G. (1984). Interactions of temperature and sublethal environmental copper exposure on the energy metabolism of bluegill, *Lepomis macrochirus* Rafinesque. *Journal of Fish Biology*, **25**, 445–53.

Fezy, J.S., Spencer, D.F. & Greene, R.W. (1979). The effect of nickel on the growth of the freshwater diatom *Navicula pelliculosa*. *Environmental Pollution*, **20**, 131–7.

Filbin, G.J. & Hough, R.A. (1979). The effects of excess copper sulphate on the metabolism of the duckweed *Lemna minor*. *Aquatic Botany*, **7**, 79–86.

Filipek, L.H., Nordstrom, D.K. & Ficklin, W.H. (1987). Interaction of acid mine drainage with waters and sediments of West Squaw Creek in the West Shasta mining district, California. *Environmental Science and Technology*, **21**, 388–96.

Florence, T.M. (1977). Trace metal species in freshwaters. *Water Research*, **11**, 681–7.

Florence, T.M. (1980). Speciation of zinc in natural waters. pp. 199–227. In: Nriagu, J.O. (Ed.) *Zinc in the Environment Part 1: Ecological Cycling*. Wiley-Interscience, New York.

Fogg, G.E. & Westlake, D.F. (1955). The importance of extracellular products of algae in freshwater. *Verhandlungen, Internationale Vereinigung für Theoretische und Angewandte Limnologie*, **12**, 219–32.

Fontaine, T.D. (1984). A non-equilibrium approach to modelling toxic metal speciation in acid, aquatic systems. *Ecological Modelling*, **22**, 85–100.

Forstner, U. & Prosi, F. (1978). Heavy metal pollution in freshwater ecosystems. pp. 129–61. In: O'Ravera, (Ed.) *Biological Aspects of Freshwater Pollution*. Pergamon Press.

Foster, P.L. (1977). Copper exclusion as a mechanism of tolerance in a green algae. *Nature (London)*, **269**, 322–3.

Foster, P.L. (1982*a*). Species associations and metal contents of algae from rivers polluted by heavy metals. *Freshwater Biology*,**12**, 17–39.

Foster, P.L. (1982*b*). Metal resistances of Chlorophyta from rivers polluted by heavy metals. *Freshwater Biology*,**12**, 41–61.

Fott, B. & McCarthy, A.J. (1964). Three acidophilic volvocine flagellates in pure culture. *Journal of Protozoology*, **11**, 116–20.

Francke, J.A. & Hillebrand, H. (1980). Effects of copper on some filamentous chlorophyta. *Aquatic Botany*, **8**, 285–9.

Frevert, T. (1985). Heavy metals in Lake Kinneret (Israel). I. Total copper and cupric ion concentrations in Lake Kinneret and the River Jordan. *Archiv für Hydrobiologie*, **104**, 527–42.

Frevert T. (1987). Heavy metals in Lake Kinneret (Israel). II. Hydrogen sulfide dependent precipitation of copper, cadmium, lead and zinc. *Archiv für Hydrobiologie*, **109**, 1–24.

Frevert, T. & Sollmann, C. (1987). Heavy metals in Lake Kinneret (Israel). III. Concentrations of iron, manganese, nickel, cobalt, molybdenum, zinc, cadmium, lead and copper in interstitial water and sediment dry weights. *Archiv für Hydrobiologie*, **109**, 181–205.

Fritz, S.C. & Carlson, R.E. (1982). Stratiographic diatom and chemical evidence for acid strip-mine lake recovery. *Water. Air & Soil Pollution*, **17**, 151–63.

Fryer, G. (1980). Acidity and species diversity in freshwater crustacean faunas. *Freshwater Biology*, **10**, 41–5.

Fujita, M. (1985). The presence of two Cd-binding components in the roots of water hyacinth cultivated in a Cd^{2+}-containing medium. *Plant and Cell Physiology*, **26**, 295–300.

Fujita, M. & Kawanishi, T. (1986). Purification and characterization of a Cd-binding complex from the root tissue of water hyacinth cultivated in a Cd^{2+}-containing medium. *Plant and Cell Physiology*, **27**, 1317–25.

Gachter, R., Davies, J.S. & Mares, A. (1978). Regulation of copper availability to phytoplankton by macromolecules in lake water. *Environmental Science and Technology*, **12**, 1416–22.

Galbraith, J.H., Williams, R.E. & Siems, P.L. (1972). Migration and leaching of metals from old mine tailings deposits. *Groundwater*, **10**, 33.

Galloway, J.N., Likens, G.E. & Edgerton, E.S. (1976). Acid precipitation in the Northeastern United States: pH and acidity. *Science (New York)*, **194**, 722–4.

Gauss, J.D., Woods, P.E., Winner, R.W. & Skillings, J.H. (1985). Acute toxicity of copper to three life stages of *Chironomus tentans* as affected by water hardness — alkalinity. *Environmental Pollution Series A,* **37**, 149–57.

Gibbs, R.J. (1973). Mechanisms of trace metal transport in rivers. *Science (New York),* **180**, 71–3.

Gibbs, R.J. (1977). Transport phases of transition metals in the Amazon and Yukon rivers. *Geological Society of America Bulletin,* **88**, 829–43.

Gibson, C.E. (1972). The algicidal effect of copper on a green and a blue-green algae and some ecological considerations. *Journal of Applied Ecology,* **9**, 513–18.

Golterman, H.L. Clymo, R.S. & Ohnstad, M.A.M. (1978). *Methods for the Physical and Chemical Analysis of Freshwaters.* 213 pp. Blackwell, Oxford.

Gopal, T., Rana, B.C. & Kumar, H.D. (1975). Autecology of the blue-green alga *Nodularia spumigena* Mertens. *Nova Hedwigia,* **26**, 225–32.

Gorham, E. (1956). On the chemical composition of some waters from the Moor House Nature Reserve. *Journal of Ecology,* **44**, 375–82.

Gorham, E. & Gordon, A.G. (1963). Some effects of smelter pollution upon aquatic vegetation near Sudbury, Ontario. *Canadian Journal of Botany,***14**, 371–8.

Gottschlich, D.E., Bell, P.R.F. & Greenfield, P.F. (1986). Estimating the rate of generation of acid drainage products in coal storage heaps. *Environmental Technology Letters,* **7**, 1–12.

Graham, G.A., Byron, G. & Norris, R.H. (1986). Survival of *Salmo gairdneri* (Rainbow Trout) in the zinc polluted Molongo River near Captains Flat, New South Wales, Australia. *Bulletin of Environmental Contamination and Toxicology,* **36**, 186–91.

Greenfield, J.P. & Ireland, M.P. (1978). A survey of the macrofauna of a coal-waste polluted Lancashire system. *Environmental Pollution,* **16**, 105–22.

Grim, E.C. & Hill, R.D. (1974). *Environmental protection in surface mining of coal.* 276 pp. EPA Publication no. EPA-670/2-74-093, National Environmental Research Center, Cincinnatti, Ohio.

Gunkel, G. & Sztraka, A. (1986). Untersuchungen zum Verhalten von Schwermetallen in Gewässern. II. Die Bedeutung der Eisen- und Mangen- Remobilisierung für die hypolimnische Anreicherung von Schwermetallen. *Archiv für Hydrobiologie,* **106**, 91–117.

Gupta, P.K., Khangarot, B.S. & Durve, V.S. (1981). The temperature dependence of the acute toxicity of copper to a freshwater pondsnail *Villiparus bengalensis*. *Hydrobiologia,* **83**, 461–4.

Gupta, S.K. & Chen, K.Y. (1975). Partitioning of trace metals in selective chemical fractions of nearshore sediments. *Environmental Letters,* **10**, 129–58.

Hale, J.G. (1977). Toxicity of metal mining wastes. *Bulletin of Environmental Contamination and Toxicology,* **17**, 66–73.

Hall, T.M. (1982). Free ionic nickel accumulation and localization in the freshwater zooplankter, *Daphnia magna*. *Limnology and Oceanography,* **27**, 718–27.

Hall, W.S., Dickson, K.L., Faleh, F., Rodgers, J.H., Wilcox, D. & Entazami, A. (1986). Effects of suspended solids on the acute toxicity of zinc to *Daphnia magna* and *Pimephales promelas*. *Water Resources Bulletin,* **22**, 913–20.

Hamilton-Taylor, J., Willis, M. & Reynolds, C.S. (1984). Depositional fluxes of metals and phytoplankton in Windermere as measured by sediment traps. *Limnology and Oceanography,* **29**, 695–710.

Hancock, F.D. (1973*a*). Algal ecology of a stream polluted through gold mining on the Witwatersrand. *Hydrobiologia,* **43**, 189–229.

Hancock, F.D. (1973*b*). The ecology of diatoms of the Klip River, Southern Transvaal. *Hydrobiologia,* **42**, 243–84.

Harding, J.P.C. & Whitton, B.A. (1976). Resistance to zinc of *Stigeoclonium tenue* in the field and the laboratory. *British Phycological Journal,* **11**, 417–26.

Harding, J.P.C. & Whitton, B.A. (1977). Environmental factors reducing the toxicity of Zn to *Stigeoclonium tenue. British Phycological Journal,* **12**, 17–21.

Harding, J.P.C. & Whitton, B.A. (1978). Zinc, cadmium and lead in water, sediments and submerged plants of the Derwent Reservoir, Northern England. *Water Research,* **12**, 307–16.

Harding, J.P.C. & Whitton, B.A. (1981). Accumulation of Zn, Cd and Pb by field populations of *Lemanea. Water Research,* **15**, 301–19.

Hargreaves, J.W. & Whitton, B.A. (1976*a*). Effect of pH on growth of acid stream algae. *British Phycological Journal,* **11**, 215–223.

Hargreaves, J.W. & Whitton, B.A. (1976*b*). Effect of pH on tolerance of *Hormidium rivulare* to zinc and copper. *Oecologia (Berlin),* **26**, 235–43.

Hargreaves, J.W., Lloyd, E.J.H. & Whitton, B.A. (1975). Chemistry and vegetation of highly acidic streams. *Freshwater Biology,* **5**, 563–76.

Harrison, A.D. (1958). The effects of sulphuric acid pollution on the biology of streams in the Transvaal, South Africa. *Verhandlungen, Internationale Vereinigung für Theoretische und Angewandte Limnologie,* **13**, 603–10.

Harrison, A.D. (1965). *Some environmental effects of coal and gold mining on the aquatic biota. Biological problems in water pollution.* Public Health Service Publication (South Africa) 999-WP-25, 270–4.

Harrison, A.D. (1984). The acidophilic thiobacilli and other acidophilic bacteria that share their habitat. *Annual Review of Microbiology,* **38**, 265–92.

Harrison, G.I., Campbell, P.G.C. & Tessier, A. (1986). Effects of pH changes on zinc uptake by *Chlamydomonas variabilis* grown in batch culture. *Canadian Journal of Fisheries and Aquatic Science,* **43**, 687–93.

Hart, B.T., Davies, S.H.R. & Thomas, P.A. (1982). Transport of iron, manganese, cadmium, copper and zinc by Magela Creek, Northern Territory, Australia. *Water Research,* **16**, 605–12.

Heath, A.G. (1984). Changes in tissue adenylates and water content of bluegill, *Lepomis macrochirus* exposed to copper. *Journal of Fish Biology,* **24**, 299–309.

Hébrard, J.-P. & Foulquier, L. (1975). Introduction à l'étude de la fixation du manganese-54 par *Platyhypnidium riparioides* (Hedw.) Dix. *Revue Bryologique et Lichenologique,* **41**, 35–54.

Hegi, H-R. & Geiger, W. (1979). Schwermetalle (Hg, Cd, Cu, Pb, Zn) in Lebern und Muskulatur des Flussbarsches (*Perca fluviatilis*) aus Bielersee und Walensee. *Schweizerische Zeitschrift für Hydrologie,* **41**, 94–107.

Heisey, R.M. & Damman, A.W.H. (1982). Copper and lead uptake by aquatic macrophytes in eastern Connecticut, U.S.A. *Aquatic Botany,* **14**, 213–29.

Hellawell, J.M. (1978). *Biological Surveillance of Rivers.* 332 pp. Water Research Centre, Medmenham.

Hem, J.D. (1972). Chemistry and occurrence of cadmium and zinc in surface water and groundwater. *Water Resources Research*, **8**, 661–79.

Henry, M.G. & Atchison, G.J. (1986). Behavioural changes in social groups in bluegills exposed to copper. *Transactions of the American Fisheries Society*, **115**, 590–5.

Herricks, E.E. & Cairns, J. (1977). The effects of lime neutralization of acid mine drainage on stream ecology. pp. 477–86. In: *Proceedings of the 32nd Industrial Waste Conference*, Purdue University.

Hiller, J.M. & Perlmutter, A. (1971). Effect of zinc on viral-host interactions in a rainbow trout cell line, RTG-2. *Water Research*, **5**, 703–10.

Hosiaisluoma, V. (1975). On the ecology of *Euglena mutabilis* on peat bogs in Finland. *Annales Botanici Fennici*, **12**, 35–6.

Howarth, R.S. & Sprague, J.B. (1978). Copper; lethality to rainbow trout in waters of various hardness and pH. *Water Research*, **12**, 455–62.

Hubschman, J.H. (1967*a*). Effects of copper on the crayfish *Orconectes rusticus*. I. Acute toxicity. *Crustaceana*, **12**, 33–42.

Hubschman, J.H. (1967*b*). Effects of copper on the crayfish *Orconectes rusticus* (Girard). II. Mode of toxic action. *Crustaceana*, **12**, 141–50.

Huckabee, J.W. (1975). Acid rock in the Great Smokies: unanticipated impact on aquatic biota of road construction in regions of sulphide mineralization. *Transactions of the American Fisheries Society*, **104**, 677–84.

Hughes, G.M. & Tort, L. (1985). Cardio-respiratory responses of Rainbow trout during recovery from zinc treatment. *Environmental Pollution Series A*, **37**, 255–66.

Hutchinson, N.J. & Sprague, J.B. (1986). Toxicity of trace metal mixtures to American Flagfish (*Jordanella floridae*) in soft, acidic water and implications for cultural acidification. *Canadian Journal of Fisheries and Aquatic Science*, **43**, 647–55.

Hutchinson, T.C. & Czyrska, H. (1975). Heavy metal toxicity and synergism to floating aquatic plants. *Verhandlungen, Internationale Vereinigung für Theoretische und Angewandte Limnologie*, **19**, 2102–11.

Hutchinson, T.C., Fedorendo, A., Fitchko, J., Kuja, A., Van Loon, J. & Lichwa, J. (1975). Movement and compartmentation of nickel and copper in an aquatic ecosystem. pp. 89–105. In: Hemphill, D.D. (Ed.) *Trace Substances in Environmental Health — IX*. University of Missouri, Columbia.

Hutchinson, T.C., Fedorendo, A., Fitchko, J., Kuja, A., Van Loon, J. & Lichwa, J. (1976). Movement and compartmentation of nickel and copper in an aquatic ecosystem. pp. 565–85. In: Nriagu, J.O. (Ed.) *Environmental Biogeochemistry*. Ann Arbor, Michigan.

Hynes, H.B.N. (1960). *The Biology of Polluted Waters*. 202 pp. Liverpool University Press, Liverpool.

Ireland, M.P. (1975). Distribution of lead, zinc and calcium in *Dendrobaena rubida* (Oligochaeta) living in soil contaminated by base metal mining in Wales. *Comparative Biochemistry and Physiology*, **52B**, 551–5.

Jackim, E. (1973). Influence of lead and other metals on fish δ-aminolevulinate dehydrase activity. *Journal of the Fisheries Research Board of Canada*, **30**, 560–2.

Jackson, T.A. (1978). A biochemical study of heavy metals in lakes and streams, and a proposed method for limiting heavy metal pollution of natural waters. *Verhandlungen, Internationale Vereinigung für Theoretische und Angewandte Limnologie*, **20**, 1945–6.

Jardim, W.F. & Pearson, H.W. (1984). A study of the copper-complexing compounds released by some species of cyanobacteria. *Water Research*, **18**, 985–9.

Jarvie, A.W.P., Markall, R.N. & Potter, H.R. (1975). Chemical alkylation of lead. *Nature (London)*, **255**, 217–18.

Jenne, E.A. (1968). Controls on Mn, Fe, Co, Ni, Cu and Zn concentrations in soils and water: the significant role of hydrous Mn and Fe oxides. *American Chemical Society Advances in Chemistry Series*, **73**, 337–87.

Jennett, J.C. & Foil, J.L. (1979). Trace metal transport from mining, milling and smelting watersheds. *Journal of the Water Pollution Control Federation*, **51**, 378–404.

Jensen, A. & Jørgensen, S.E. (1984). Analytical chemistry applied to metal ions in the environment. pp. 5-59. In: Sigel, H. (Ed.) *Metal Ions in Biological Systems*. Marcel Dekker Inc., New York & Basel.

Jensen, T.E., Rachlin, J.W., Jani, V. & Warkentine, B.E. (1986). Heavy metal uptake in relation to phosphorus nutrition in *Anabaena variabilis* (Cyanophyceae). *Environmental Pollution Series A*, **42**, 261–71.

Johnson, C.A. (1986). The regulation of trace element concentrations in river and estuarine waters contaminated with acid mine drainage: the adsorption of Cu and Zn on amorphous Fe oxyhydroxides. *Geochimica et Cosmochimica Acta*, **50**, 2433–8.

Johnson, C.A. & Thornton, I. (1987). Hydrological and chemical factors controlling the concentrations of Fe, Cu, Zn and As in a river system contaminated with acid mine drainage. *Water Research*, **21**, 359–65.

Johnson, M.G. (1987). Trace element loadings to sediments of fourteen Ontario Lakes and correlations with concentrations in fish. *Canadian Journal of Fisheries and Aquatic Sciences*, **44**, 3–13.

Jones, A.N. & Howells, W.R. (1975). The partial recovery of the metal polluted River Rheidol. pp. 443–59. In: Chadwick, M.J. & Goodman, G.T. (Eds) *The Ecology of Resource Degradation and Renewal*. Symposium of the British Ecological Society No. 15, Blackwell, Oxford.

Jones, J.R.E. (1940). A study of the zinc-polluted River Ystwyth in North Cardiganshire, Wales. *Annals of Applied Biology*, **27**, 366–78.

Jones, J.R.E. (1949). An ecological study of the River Rheidol, North Cardiganshire, Wales. *Journal of Animal Ecology*, **18**, 67–88.

Jones, J.R.E. (1958). A further study of the zinc-polluted River Ystwyth. *Journal of Animal Ecology*, **27**, 1–14.

Jones, K.C., Peterson, P.J. & Davies, B.E. (1985). Silver and other metals in some aquatic bryophytes from streams in the lead mining district of Mid-Wales, Great Britain. *Water, Soil & Air Pollution*, **24**, 329–38.

Jørgensen, S.E. & Jensen, A. (1984). Processes of metal ions in the environment. pp. 61–103. In: Sigel, H. (Ed.) *Metal Ions in Biological Systems*. Marcel Dekker Inc. New York & Basel.

Judy, R.D. (1979). The acute toxicity of copper to *Gammarus fasciatus*, a freshwater amphipod. *Bulletin of Environmental Contamination and Toxicology*, **21**, 219–24.

Kashyap, A.K. & Gupta, S.L. (1982). Effect of lethal copper concentrations on nitrate uptake, reduction and nitrate release in *Anacystis nidulans*. *Zeitschrift für Pflanzenphysiologie*, **107**, 289–94.

Keeney, W.L., Breck, W.G., Van Loon, G.W. & Page, J.A. (1976). The determination of trace metals in *Cladophora glomerata* — *C. glomerata* as a potential biological monitor. *Water Research*, **10**, 981–4.

Kelly, M.G. (1986). Heavy metals and aquatic bryophytes: accumulation and their use as monitors. 304 pp. PhD thesis, University of Durham, England.

Kelly, M.G. & Whitton, B.A. (1988). Interspecific differences in Zn, Cd and Pb accumulation by freshwater algae and bryophytes. *Hydrobiologia*, in press.

Kelly, M.G., Girton, C. & Whitton, B.A. (1987). Use of moss-bags for monitoring heavy metals in rivers. *Water Research*, **21**, 1429–35.

Khangarot, B.S. & Ray, P.K. (1987*a*). Studies on the acute toxicity of copper and mercury alone and in combination to the common guppy *Poecilia reticulata* (Peters). *Archiv für Hydrobiologie*, **110**, 303–14.

Khangarot, B.S. & Ray, P.K. (1987*b*). Zinc sensitivity of a freshwater snail, *Lymnaea luteola* L., in relation to seasonal variations in temperature. *Bulletin of Environmental Contamination and Toxicology*, **39**, 45–9.

King, D.L., Simmler, J.J., Decker, C.S. & Ogg, C.W. (1974). Acid strip mine lake recovery. *Journal of the Water Pollution Control Federation*, **46**, 2301–15.

Kinsman, D.J.J. (1984). Ecological effects of deposited S and N compounds: effects on aquatic biota. *Philosophical Transactions of the Royal Society of London Series B*, **305**, 479–84.

Klotz, R.L. (1981). Algal response to copper under riverine conditions. *Environmental Pollution Series A*, **24**, 1–19.

Koryak, M., Shapiro, M.A. & Sykora, J.L. (1972). Riffle zoobenthos in streams receiving acid mine drainage. *Water Research*, **6**, 1239–47.

Kostyayev, V.Ya, Yagodka, S.N. & Sokolov, V.A. (1981). Sensitivity of *Anabaena spiroides* to zinc and cobalt. *Gidrobiologicheskii Zhurnal*, **16**, 81–4.

Kotangale, L.R., Sarkar, R. & Krishnamoorthi, K.P. (1984). Toxicity of mercury and zinc to *Spirulina platensis*. *Indian Journal of Environmental Health*, **26**, 41–6.

Kuwabara, J.S. (1985). Phosphorus-zinc interactive effects on growth by *Selanastrum capricornutum* (Chlorophyta). *Environmental Science and Technology*, **19**, 417–21.

Kuwabara, J.S. & Leland, H.V. (1986). Adaptation of *Selanastrum capricornutum* (Chlorophyceae) to copper. *Environmental Toxicology and Chemistry*, **5**, 197–203.

Kuwabara, J.S., Leland, H.V. & Bencala, K.E. (1984). Copper transport along a Sierra Nevada stream. *Journal of Environmental Engineering* **110**, 646–55.

Lackey, J.B. (1938). The flora and fauna of surface waters polluted by acid mine drainage. *Public Health Reports*, **53**, 1499–1507.

Lampkin, A.J. & Sommerfield, M.R. (1982). Algal distribution in a small intermittent stream receiving acid mine drainage. *Journal of Phycology*, **18**, 196–9.

Lampkin, A.J. & Sommerfield, M.R. (1986). Impact of acid mine-drainage from abandoned spoils on the chemistry of an intermittent stream in the arid Southwest. *Hydrobiologia*, **139**, 135–42.

Lane, A.E. & Burris, J.E. (1981). Effects of environmental pH on the internal pH of *Chlorella pyrenoidosa*, *Scenedesmus quadricauda* and *Euglena mutabilis*. *Plant Physiology, Lancaster*, **68**, 439–42.

LaPerriere, J.D., Wagener, S.M. & Bjerklie, D.M. (1985). Gold mining effects on heavy metals in streams, Circle Quadrangle, Alaska. *Water Resources Bulletin*, **21**, 245–52.

LaPoint, T.W., Melancon, S.M. & Morris, M.K. (1984). Relationships among observed metal concentrations, criteria and benthic community structural responses in 15 streams. *Journal of the Water Pollution Control Federation*, **56**, 1030–8.

Laube, V.M., McKenzie, C.N. & Kushner, D.J. (1980). Strategies of response to copper, cadmium, and lead by a blue-green and a green alga. *Canadian Journal of Microbiology*, **26**, 1300–11.

Laurén, D.J. & McDonald, D.G. (1986). Influence of water hardness, pH, and alkalinity on the mechanisms of copper toxicity in juvenile Rainbow Trout, *Salmo gairdneri*. *Canadian Journal of Fisheries and Aquatic Science*, **43**, 1488–96.

Laurén, D.J. & McDonald, D.G. (1987*a*). Acclimation to copper by rainbow trout, *Salmo gairdneri*: physiology. *Canadian Journal of Fisheries and Aquatic Science*, **44**, 99–104.

Laurén, D.J. & McDonald, D.G. (1987*b*). Acclimation to copper by rainbow trout, *Salmo gairdneri*: biochemistry. *Canadian Journal of Fisheries and Aquatic Science*, **44**, 105–11.

Laurie, R.D. & Jones, J.R.E. (1938). The faunistic recovery of a lead-polluted river in North Cardiganshire, Wales. *Journal of Animal Ecology*, **7**, 272–89.

Laxen, D.P.H. (1984). Adsorption of cadmium, lead and copper during the precipitation of hydrous ferric oxide in a natural water. *Chemical Geology*, **47**, 321–32.

Laxen, D.P.H. & Harrison, R.M. (1983). Physico-chemical speciation of selected metals in the treated effluent of a lead-acid battery manufacturer and in the receiving water. *Water Research*, **17**, 71–80.

Lazareva, L.P. (1986). Changes in biological characteristics of *Daphnia magna* from chronic action of copper and nickel at low concentrations. *Hydrobiology Journal*, **21**(5), 59–62.

LeBlanc, G.A. (1982). Laboratory investigation into the development of resistance of *Daphnia magna* (Strauss) to environmental pollutants. *Environmental Pollution Series A*, **27**, 309–22.

LeBlanc, G.A. (1985). Effects of copper on the competitive interactions of two species of Cladocera. *Environmental Pollution Series A*, **37**, 13–25.

Leckie, J.O. & Davies, J.A. (1979). Aqueous environmental chemistry of copper. pp. 90–121. In: Nriagu, J.O. (Ed.) *Copper in the Environment Part 1: Ecological Cycling*. Wiley-Interscience, New York.

Leivestad, H. & Muniz, I.P. (1976). Fish kills at low pH in a Norwegian river. *Nature (London)*, **259**, 391–2.

Leland, H.V. & Carter, J.J. (1984). Effects of copper on species composition of periphyton in a Sierra Nevada, California, stream. *Freshwater Biology,* **14**, 281–96.

Les, A. & Walker, R.W. (1984). Toxicity and binding of copper, zinc, and cadmium by the blue-green alga, *Chroococcus paris. Water, Air & Soil Pollution,* **23**, 129–39.

Letterman, R.D. & Mitsch, W.J. (1978). Impact of mine drainage on a mountain stream in Pennsylvania. *Environmental Pollution,* **17**, 53–74.

Lion, L.W., Altmann, R.S. & Leckie, J.O. (1982). Trace-metal adsorption characteristics of estuarine particulate matter: evaluation of contributions of Fe/Mn oxide and organic surface coatings. *Environmental Science and Technology,* **16**, 660–6.

Lloyd, R. (1960). The toxicity of zinc sulphate to rainbow trout. *Annals of Applied Biology,* **48**, 84–94.

Lloyd, R. (1961). The toxicity of mixtures of zinc and copper sulphate to rainbow trout (*Salmo gairdneri* Richardson). *Annals of Applied Biology,* **49**, 535–8.

Lloyd, R. & Jordan, D.H.M. (1964). Some factors affecting the resistance of rainbow trout (*Salmo gairdneri* Richardson) to acid waters. *International Journal of Air and Water Pollution,* **8**, 393–403.

Lorenz, M.G. & Krumbein, W.E. (1984). Large-scale determination of cyanobacterial susceptibility to antibiotics and inorganic ions. *Applied Microbiology and Biotechnology,* **20**, 422–6.

Lovell, H.L. (1983). Coal mine drainage in the United States — an overview. pp. 1–25. In: Odendaal, P.E. (Ed.) *Mine Water Pollution Proceeds — International Association of Water Pollution Research.* Water Science and Technology (South Africa) **15**(2), 1–181.

Lundgren, D.G. & Silver, M. (1980). Ore leaching by bacteria. *Annual Review of Microbiology,* **34**, 263–83.

Mackay, R.J. & Kersey, K.E. (1985). A preliminary study of aquatic insect communities and leaf decomposition in acid streams near Dorset, Ontario. *Hydrobiologia,* **122**, 3–11.

Malacea, I. & Gruia, E. (1965). Contributii a cunoasterea actuinii toxice a cuprului, zincului, plumbului si nichelului asupra unor specii di pesti si a dafnei. *Studii de Protectia si Epurarea Apelor,* **6**, 391–451.

Malanchuk, J.L. & Gruendling, G.K. (1973). Toxicity of lead nitrate to algae. *Water, Air & Soil Pollution,* **2**, 181–90.

Mal Reddy, N. & Venkateswara Rao, P. (1987). Copper toxicity to the freshwater snail, *Lymnaea luteola. Bulletin of Environmental Contamination and Toxicology,* **39**, 50–5.

Mance, G. (1987). *Pollution Threat of Heavy Metals in the Aquatic Environment.* 372 pp. Elsevier Applied Science Publishers, London & New York.

Manly, R. & George, W.O. (1977). The occurrence of some heavy metals in populations of the freshwater mussel *Anodonta anatina* (L.) from the River Thames. *Environmental Pollution,* **14**, 139–54.

Mantoura, R.F.C., Dickson, A. & Riley, R.P. (1978). The complexation of metal, with humic materials in natural waters. *Estuarine and Coastal Marine Sciences,* **6**, 387–408.

Markarian, R.K., Mathews, M.C. & Conner, L.T. (1980). Toxicity of nickel, copper, zinc and aluminium mixtures to the white sucker (*Catostomus commersoni*). *Bulletin of Environmental Contamination and Toxicology,* **25**, 790–6.

Marquenie van der Werff, M. & Pruyt, M.J. (1982). Long term effects of heavy metals on aquatic plants. *Chemosphere*, **11**, 727–39.

Martin, J.H., Knauer, G.A. & Flegal, A.R. (1980). Distribution of zinc in natural waters. pp. 193–7. In: Nriagu, J.O. (Ed.) *Zinc in the Environment Part I: Ecological Cycling*. Wiley-Interscience, New York.

Mathis, B.J. & Cummings, T.F. (1973). Selected metals in sediments, water and biota in the Illionois River. *Journal of the Water Pollution Control Federation*, **45**, 1573–83.

Mathis, B.J. & Kevern, N.R. (1975). Distribution of mercury, cadmium, lead and thallium in a eutrophic lake. *Hydrobiologia*, **46**, 207–22.

Matter, W.J., Ney, J.J. & Maughan, O.E. (1978). Sustained impact of abandoned surface mines on fish and benthic invertebrate populations in headwater streams of Southwestern Virginia. pp. 203–16. In: *Surface Mining and Fish/Wildlife Needs in Eastern U.S.A.*. Symposium, West Virginia, 3–6 December 1978. US Fish and Wildlife Service.

McBrien, D.C.H. & Hassall, K.A. (1965). Loss of cell potassium by *Chlorella vulgaris* after contact with toxic amounts of copper sulphate. *Physiologia Plantarum*, **18**, 1059–65.

McDuffie, B., El-Barbary, I., Holland, G.J. & Tiberio, R.D. (1977). Heavy metals in Susquehenna River bottom sediments - surficial concentrations, urban impacts and transport mechanism. pp. 467–8. In: Hutzinger, O., Van Lelyveld, I.H. & Zoeteman, B.C.J. (Eds) *Aquatic Pollutants: Transformation and Biological Effects*. Pergamon Press, Oxford.

McIntosh, A.W. & Keveren, N.R. (1974). Toxicity of copper to zooplankton. *Journal of Environmental Quality*, **3**, 166–70.

McKim, J.M., Eaton, J.G. & Holcombe, G.W. (1978). Metal toxicity to embryos and larvae of eight species of freshwater fish. II. copper. *Bulletin of Environmental Contamination and Toxicology*, **19**, 608–16.

McLean, R.O. & Jones, A.K. (1975). Studies of tolerance to heavy metals in the flora of the rivers Ystwyth and Clarach, Wales. *Freshwater Biology*, **5**, 431–44.

McLeay, D.J., Birtwell, I.K., Hartman, G.F. & Ennis, G.L. (1987). Responses of Arctic grayling (*Thymallus arcticus*) to acute and prolonged exposure to Yukon placer mining sediment. *Canadian Journal of Fisheries and Aquatic Science*, **44**, 658–73.

McMurtry, M.J. (1984). Avoidance of sublethal doses of copper and zinc by tubificid oligochaetes. *Journal of Great Lakes Research*, **10**, 267–72.

McNaughton, S.J., Folsom, T.C., Lee, T., Pork, F., Price, C., Roeder, D., Schmitz, J. & Stockwell, C. (1974). Heavy metal tolerance in *Typha latifolia* without the evolution of tolerant races. *Ecology*, **55**, 1163–5.

McNight, D.M. & Feder, G.L. (1984). The ecological effects of acid conditions and precipitation of hydrous metal oxides in a Rocky Mountain stream. *Hydrobiologia*, **119**, 129–38.

McNight, D.M. & Morel, F.M.M. (1980). Copper complexation by siderophores from filamentous blue-green algae. *Limnology and Oceanography*, **25**, 62–71.

Merlini, M. & Pozzi, G. (1977). Lead and freshwater fishes. Part 1 — Lead accumulation and water pH. *Environmental Pollution*, **12**, 167–72.

Miller, G.E., Wile, I. & Hitchin, G.G. (1983). Patterns of accumulation of selected metals in members of the softwater macrophyte flora of the central Ontario lakes. *Aquatic Botany*, **15**, 53–64.

Millero, F.J. (1975). The physical chemistry of estuaries. pp. 25–55. In: Church, T. (Ed.) *Marine Chemistry in the Coastal Environment*. ACS Symposium Series 18.

Mink, L.L., Williams, R.E. & Wallace, A.T. (1972). Effects of early day mining operations on present day water quality. *Groundwater*, **10**, 17–26.

Monahan, T.J. (1973). Lead inhibition of *Hormotila blennista* (Chlorophyceae, Chlorococcales). *Phycologia*, **12**, 247.

Monahan, T.J. (1976). Lead inhibition of Chlorophycean microalgae. *Journal of Phycology*, **12**, 358–62.

Moon, T.C. & Lucostic, C.M. (1979). Effects of acid mine drainage on a Southwestern Pennsylvania stream. *Water, Air & Soil Pollution*, **11**, 377–90.

Moore, J.W. & Ramamoorthy, S. (1984). *Heavy Metals in Natural Waters. Applied Monitoring and Impact Assessment*. 268 pp. Springer-Verlag, New York.

Morel, F.M.M., McDuff, R.G. & Morgan, J.J. (1973). Interactions and chemostasis in aquatic chemical systems: role of pH, *pε*, solubility and complexation. pp. 157–200. In: Singer, P.C. (Ed.) *Trace Metals and Metal-Organic Interactions in Natural Waters*. Ann Arbor Science, Ann Arbor.

Moriarty, F., Bull, K.R., Hansom, H.M. & Freestone, P. (1982). The distribution of Pb, Zn and Cd in sediments of an ore-enriched lotic ecosystem, the River Ecclesbourne, Derbyshire. *Environmental Pollution Series B*, **4**, 45–68.

Moriarty, F., Hanson, H.M. & Freestone, P. (1984). Limitations of body burden as an index of environmental contamination: heavy metals in fish *Cottus gobio* L. from the River Ecclesbourne, Derbyshire. *Environmental Pollution Series A*, **34**, 297–320.

Mount, D.I. (1966). The effect of total hardness and pH on acute toxicity of zinc to fish. *International Journal of Air and Water Pollution*, **10**, 49–56.

Mount, D.I. (1968). Chronic toxicity of copper to fathead minnows (*Pimephales promelas*, Rafinesque). *Water Research*, **2**, 215–23.

Mouvet, C. & Bourg, A.C.M. (1983). Speciation (including adsorbed species) of copper, lead, nickel and zinc in the Meuse River. *Water Research*, **17**, 641–9.

Mudroch, A. & Capobianco, J.A. (1980). Impact of past mining activities on aquatic sediments in Moira river basin, Ontario. *Journal of Great Lakes Research*, **6**, 121–8.

Mukherji, S. & Maitra, P. (1976). Toxic effects of lead on growth and metabolism of germinating rice (*Oryza sativa* L.) seeds and on mitosis of onion (*Allium cepa* L.) root tip cells. *Indian Journal of Experimental Biology*, **14**, 519–21.

Mukherji, S. & Maitra, P. (1977) Growth and metabolism of germinating rice (*Oryza sativa* L.) seeds as influenced by toxic concentrations of lead. *Zeitschrift für Pflanzenphysiologie*, **81**, 26–33.

Murphy, B.R., Atchinson, G.J. & McIntosh, A.W. (1978). Cd and Zn in muscle of bluegill (*Lepomis macrochirus*) and largemouth Bass (*Micropterus salmoides*) from an industrially contaminated lake. *Environmental Pollution*, **17**, 253–7.

Murti, R. & Shukla, G.S. (1984). Toxicity of copper sulphate and zinc sulphate to *Macrobachium lamarrei* (H. Milne Edwards) (Decapoda, Palaemonidae). *Crustaceana*, **47**, 168–73.

Nasu, Y. & Kugimoto, M. (1981). *Lemna* (duckweed) as an indicator of water pollution. I. The sensitivity of *Lemna paucicostata* to heavy metals. *Archives of Environmental Contamination and Toxicology*, **10**, 159–69.

Nasu, Y., Kugimoto, M., Tanaka, O. & Takimoto, A. (1984*a*). *Lemna* as an indicator of water pollution and the absorption of heavy metals by *Lemna*. pp. 113–20. In: Pascoe, D. & Edwards, R.W. (Eds) *Freshwater Biological Monitoring*. Pergamon Press, Oxford.

Nasu, Y., Kugimoto, M., Tanaka, O., Yanase, D. & Takimoto, A. (1984*b*). Effects of cadmium and copper co-existing in the medium on the growth and flowering of *Lemna paucicostata* in relation to their absorption. *Environmental Pollution Series A*, **33**, 267–74.

Nehring, R.B. (1976). Aquatic insects as biological monitors of heavy metal pollution. *Bulletin of Environmental Contamination and Toxicology*, **15**, 147–54.

Newman, M.C. & McIntosh, A.W. (1982). The influence of lead in components of a freshwater ecosystem on molluscan tissue lead concentrations. *Aquatic Toxicology*, **2**, 1–19.

Nicola Giudici, M. De, Migliore, L. & Guarino, S.M. (1987). Sensitivity of *Asellus aquaticus* (L.) and *Proasellus coxalis* Dollf. (Crustacea, Isopoda) to copper. *Hydrobiologia*, **146**, 63–9.

Nieboer, E. & Richardson, D.H.S. (1980). The replacement of the nondescript term 'heavy metals' by a biologically and chemically significant classification of metal ions. *Environmental Pollution Series B*, **1**, 3–26.

Nienke, G.E. & Lee, G.F. (1982). Sorption of zinc by Lake Michigan sediments. Implications for zinc water quality criteria standards. *Water Research*, **16**, 1373–8.

Noike, T., Nakamura, K. & Matsumoto, J. (1983). Oxidation of ferrous iron by acidophilic iron-oxidizing bacteria from a stream receiving acid mine waste. *Water Research*, **17**, 21–7.

Norris, R.H. (1986). Mine waste pollution of the Molongo River, New South Wales and the Australian Capital Territory: effectiveness of remedial works at Captains Flat Mining Area. *Australian Journal of Marine and Freshwater Research*, **37**, 147–57.

Norris, R.H. & Lake, P.S. (1984). Trace-metal concentrations in fish from the South Esk River, Northeastern Tasmania. *Bulletin of Environmental Contamination and Toxicology*, **33**, 348–54.

Norris, R.H., Swain, R. & Lake, P.S. (1981). Ecological effects of mine effluents on the South Esk River, North-eastern Tasmania. III. Benthic macroinvertebrates. *Australian Journal of Marine and Freshwater Research*, **33**, 789–809.

Norris, R.H., Lake, P.S. & Swain, R. (1982). Ecological effects of mine effluents on the South Esk River, North-eastern Tasmania. II. Trace metals. *Australian Journal of Marine and Freshwater Research*, **32**, 165–73.

Nriagu, J.O., Wong, H.K.T. & Cocker, R.D. (1981). Particulate and dissolved trace metals in Lake Ontario. *Water Research*, **15**, 91–6.

Nuttall, P.M. & Bielby, G.H. (1973). The effect of china-clay wastes on stream invertebrates. *Environmental Pollution*, **5**, 77–86.

Ochsenbein, U., Davison, W., Hilton, J. & Haworth, E.Y. (1983). The geochemical record of major cations and trace metals in a productive lake. *Archiv für Hydrobiologie*, **98**, 463–88.

O'Grady, K.T. & Abdullah, M.I. (1985). Mobility and residence of Zn in brown trout *Salmo trutta*: results of environmentally induced change through transfer. *Environmental Pollution Series A*, **38**, 109–27.

Olem, H. (1981). Coal and coal mine drainage (literature review). *Journal of the Water Pollution Control Federation*, **53**, 814–24.

Orsatti, S.D. & Colgan, P.W. (1987). Effects of sulphuric acid exposure on the behaviour of largemouth bass, *Micropterus salmoides*. *Environmental Biology of Fish*, **19**, 119–29.

Pagenkopf, G.F. & Neuman, D.R. (1974). Lead concentrations in native trout. *Bulletin of Environmental Contamination and Toxicology*, **12**, 70–5.

Pagenkopf, G.F., Russo, R.C. & Thurston, R.V. (1974). Effect of complexation on toxicity of copper to fishes. *Journal of the Fisheries Research Board of Canada*, **31**, 462–5.

Pain, S. (1987). After the goldrush. *New Scientist*, **1574**, 36–40.

Parry, G.D.R. & Hayward, J. (1973). The uptake of ^{65}Zn by *Dunaliella tertiolecta* Butcher. *Journal of the Marine Biological Association of the United Kingdom*, **53**, 915–22.

Parsons, J.W. (1952). A biological approach to the study and control of acid mine pollution. *Journal of the Tennessee Academy of Science*, **27**, 304–9.

Parsons, J.D. (1968). The effects of acid strip-mine effluents on the ecology of a stream. *Archiv für Hydrobiologie*, **65**, 25–50.

Parsons, J.D. (1977). Effects of acid mine wastes on aquatic ecosystems. *Water, Air, & Soil Pollution*, **7**, 333–54.

Patrick, R., Cairns, J. & Scheier, A. (1968). The relative sensitivity of diatoms, snails and fish to twenty constituents of industrial wastes. *Progressive Fish Culturist*, July 1968, 137–40.

Patterson, G. & Whitton, B.A. (1981). Chemistry of water, sediments and algal filaments in groundwater draining an old lead-zinc mine. pp. 65–72. In: Say, P.J. & Whitton, B.A. (Eds) *Heavy Metals in Northern England: Environmental and Biological Aspects*. Department of Botany, University of Durham, England.

Pearson, R.G. (1968). Hard and soft acids and bases, HSAB, part 1: fundamental principles. *Journal of Chemical Education*, **45**, 581–7.

Pentecost, A. (1982). The distribution of *Euglena mutabilis* in Sphagna, with reference to the Malham Tarn North Fen. *Field Studies*, **5**, 365–87.

Peterson, H.G. & Healey, F.P. (1985). Comparative pH dependent metal inhibition of nutrient uptake by *Scenedesmus quadricauda* (Chlorophyceae). *Journal of Phycology*, **21**, 217–22.

Peterson, H.G., Healey, F.P. & Wagemann, R. (1984). Metal toxicity to algae: a highly pH dependent phenomenon. *Canadian Journal of Fisheries and Aquatic Science*, **41**, 974–9.

Petersen, R. (1982). Influence of copper and zinc on the growth of a freshwater alga, *Scenedesmus quadricauda*: the significance of chemical speciation. *Environmental Science and Technology*, **16**, 443–7.

Phillips, D.J.H. (1979). *Quantitative Aquatic Biological Indicators. Their Use to Monitor Trace Metal and Organochlorine Pollution*. 488 pp. Applied Science Publishers, London.

Pickering, D.C. & Puia, I.L. (1969). Mechanism for the uptake of Zn by *Fontinalis antipyretica*. *Physiologia Plantarum*, **22**, 653–61.

Powlesland, C. & George, J. (1986). Acute and chronic toxicity of nickel to larvae of *Chironomus riparis* (Meigen). *Environmental Pollution Series A*, **42**, 47–64.

Rabe, R., Schuster, H. & Kohler, A. (1982). Effects of copper chelate on photosynthesis and some enzyme activities of *Elodea canadensis*. *Aquatic Botany*, **14**, 167–75.

Rachlin, J.W. & Farran, M. (1974). Growth response of the green alga *Chlorella vulgaris* to selective concentrations of zinc. *Water Research*, **8**, 575–7.

Rachlin, J.W., Jensen, T.E. & Warkentine, B. (1983). The growth response of the diatom *Navicula incerta* to selected concentrations of the metals cadmium, copper, lead and zinc. *Bulletin of the Torry Botanical Club*, **100**, 217–23.

Rachlin, J.W., Jensen, T.E. & Warkentine, B. (1985). Morphometric analysis of the response of *Anabaena flos-aquae* and *Anabaena variabilis* (Cyanophyceae) to selected concentrations of zinc. *Archives of Environmental Contamination and Toxicology*, **14**, 395–402.

Rai, L.C. & Kumar, A. (1980). Effects of certain environmental factors on the toxicity of zinc to *Chlorella vulgaris*. *Microbios Letters*, **13**, 79–84.

Rai, L.C., Gaur, J.P. & Kumar, H.D. (1981). Protective effects of certain environmental factors on the toxicity of zinc, mercury, and methylmercury to *Chlorella vulgaris*. *Environmental Research*, **25**, 250–9.

Rainbow, P.S. & Moore, P.G. (1986). Comparative metal analyses in amphipod crustaceans. *Hydrobiologia*, **141**, 273–89.

Raistrick, A. & Jennings, B. (1965). *A History of Lead Mining in the Pennines*. 347 pp. Longmans, London.

Ramamoorthy, S. & Rust, B.R. (1978). Heavy metal exchange processes in sediment-water systems. *Environmental Geology*, **2**, 165–72.

Rana, B.C. & Kumar, H.D. (1974). Effects of toxic waste and waste water components on algae. *Phykos*, **13**, 67–83.

Rasmussen, L. & Sand-Jensen, K. (1979). Heavy metals in acid streams from lignite mining areas. *Science of the Total Environment*, **12**, 61–74.

Ray, S.N. & White, W.J. (1979). *Equisetum arvense* — an aquatic vascular plant as a biological monitor for heavy metal pollution. *Chemosphere*, **3**, 125–8.

Rehwoldt, R., Bida, G. & Nerne, B. (1971). Acute toxicity of copper, nickel and zinc ions to some Hudson river fish species. *Bulletin of Environmental Contamination and Toxicology*, **6**, 445–8.

Rehwoldt, R., Lasko, L., Shaw, C. & Wirhowski, E. (1973). The acute toxicity of some heavy metal ions towards benthic organisms. *Bulletin of Environmental Contamination and Toxicology*, **10**, 291–4.

Reynolds, C.S. (1984). *The Ecology of Freshwater Phytoplankton*. 384 pp. Cambridge University Press, Cambridge.

Reynoldson, T.B. (1987). Interactions between sediment contaminants and benthic organisms. *Hydrobiologia*, **149**, 53–66.

Richter, R.O. & Theis, T.L. (1980). Nickel speciation in a soil/water system. pp. 189–202. In: Nriagu, J.O. (Ed.) *Nickel in the Environment*. Wiley-Interscience, New York.

Riley, C.V. (1960). The ecology of water areas associated with coal strip-mined lands in Ohio. *Ohio Journal of Science*, **60**, 106–21.

Riley, J.P. & Chester, R. (1971). *Introduction to Marine Chemistry*. 465 pp. Academic Press, London.

Ripley, E.A., Redmann, R.E. & Maxwell, J. (1979). *Environmental Impact of Mining in Canada*. 284 pp. Centre for Resources Studies, Queens University, Kingston, Ontario.

Rippey, B. (1982). Sediment–water interactions of Cu, Zn and Pb discharged from a domestic waste water source into a bay of Lough Neagh, Northern Ireland. *Environmental Pollution Series B*, **3**, 199–214.

Rippka, R., Derulles, J., Waterbury, J.B., Herdman, M. & Stanier, R.Y. (1979). Generic assignments, strain histories and properties of pure cultures of cyanobacteria. *Journal of General Microbiology*, **111**, 1–61.

Roback, S.S. & Richardson, J.W. (1969). The effect of acid mine drainage on aquatic insects. *Proceedings of the Academy of Natural Sciences of Philadelphia*, **121**, 81–107.

Robinson, P.K. & Hawkes, H.A. (1986). Studies on the growth of *Cladophora glomerata* in laboratory continuous-flow culture. *British Phycological Journal*, **21**, 437–44.

Roch, M. & McCarter, J.A. (1984). Metallothionein induction, growth and survival of chinook salmon exposed to zinc, copper, and cadmium. *Bulletin of Environmental Contamination and Toxicology*, **32**, 478–85.

Roch, M., Noonan, P. & McCarter, J.A. (1986). Determination of no effect levels of heavy metals for rainbow trout using hepatic metallothionein. *Water Research*, **20**, 771–4.

Romano, R.R. (1976). Fluvial transport of selected heavy metals in the Grand Calumet River system. pp. 207–16. In: Hemphill, D.D. (Ed.) *Trace Substances in Environmental Health — X*. University of Missouri, Columbia.

Rueter, J.G., O'Reilley, K.T. & Petersen, R.R. (1987). Indirect aluminium toxicity to the green alga *Scenedesmus* through increased cupric ion activity. *Environmental Science and Technology*, **21**, 435–8.

Sakai, H., Kojima, Y. & Saito, K. (1986). Distribution of heavy metals in water and sieved sediments in the Toyohira River. *Water Research*, **20**, 559–67.

Salomons, W., de Rooij, N.M., Kerdijk, H. & Bril, J. (1987). Sediments as a source for contaminants? *Hydrobiologia*, **149**, 13–30.

Sand, W. (1985). The influence of four detergents on the substrate oxidation by *Thiobacillus ferrooxidans*. *Environmental Technology Letters*, **6**, 439–44.

Sand-Jensen, K. & Rasmussen, L. (1978). Macrophytes and chemistry of acidic streams from lignite mining areas. *Botanisk Tidsskrift*, **72**, 105–12.

Satake, K. & Miyasaka, K. (1984). Evidence of high mercury accumulation in the cell wall of the liverwort *Jungermannia vulcanicola* Steph. to form particles of a mercury–sulphur compound. *Journal of Bryology*, **13**, 101–5.

Satake, K. & Saijo, Y. (1974). Carbon dioxide content and metabolic activity of microorganisms in some acid lakes in Japan. *Limnology and Oceanography*, **19**, 331–8.

Satake, K. Shimizu, H. & Nishikawa, M. (1984). Elemental composition of the aquatic liverwort *Jungermannia vulcanicola* Steph. in acid streams. *Journal of the Hattori Botanical Laboratory*, **56**, 241–8.

Satake, K., Nishikawa, M. & Shibata, K. (1987). Elemental composition of the sub-aquatic liverwort *Pellia endiviifolia* (Dicks.) Dum. in relation to heavy metal contamination. *Hydrobiologia*, **148**, 131–6.

Say, P.J. & Whitton, B.A. (1977). Influence of zinc on lotic plants. II. Environmental effects of toxicity of zinc to *Hormidium rivulare*. *Freshwater Biology*, **7**, 377–84.

Say, P.J. & Whitton, B.A. (1980). Changes in flora down a stream showing a zinc gradient. *Hydrobiologia*, **76**, 255–62.

Say, P.J. & Whitton, B.A. (1981). Chemistry and plant ecology of zinc-rich streams in the Northern Pennines. pp. 55–64. In: Say, P.J. & Whitton, B.A. (Eds) *Heavy Metals in Northern England: Environmental and Biological Aspects*. Department of Botany, University of Durham, England.

Say, P.J. & Whitton, B.A. (1982). Chemistry and plant ecology of zinc-rich streams in France. 2. The Pyrénées. *Annales de Limnologie*, **18**, 19–31.

Say, P.J. & Whitton, B.A. (1983). Accumulation of heavy metals by aquatic mosses. 1. *Fontinalis antipyretica*. *Hydrobiologia*, **100**, 245–60.

Say, P.J., Diaz, B.M. & Whitton, B.A. (1977). Influence of zinc on lotic plants. I. Tolerance of *Hormidium* species to zinc. *Freshwater Biology*, **7**, 357–76.

Say, P.J., Harding, J.P.C. & Whitton, B.A. (1981). Aquatic mosses as monitors of heavy metal contamination in the River Etherow, Great Britain. *Environmental Pollution Series B*, **2**, 295–307.

Schecher, W.D. & Driscoll, C.T. (1985). Interactions of copper and lead with *Nostoc muscorum*. *Water, Air & Soil Pollution*, **24**, 85–101.

Schlick, D.P. & Wahler, W.A. (1976). *Mine Refuse Impoundment in the United States*. US Department of the Interior, Mining Enforcement and Safety Administration Information Report 1028.

Schmidt, U. & Huber, F. (1976). Methylation of organolead and lead(II) compounds to $(CH_3)_4Pb$ by microorganisms. *Nature (London)*, **259**, 157–8.

Schreinemakers, W.A.C. & Dorhout R. (1985). Effects of copper ions as growth and ion absorption by *Spirodela polyrhiza* (L.) Schleiden. *Journal of Plant Physiology*, **121**, 343–51.

Scullion, J. & Edwards, R.W. (1980). The effects of coal industry pollutants on the macroinvertebrate fauna of a small river in the South Wales coalfield. *Freshwater Biology*, **10**, 141–62.

Segner, H. (1987). Response of fed and starved roach, *Rutilus rutilus*, to sublethal copper contamination. *Journal of Fish Biology*, **30**, 423–37.

Sehgal, R. & Saxena, A.B. (1986). Toxicity of zinc to a viviparous fish, *Lebistes reticulatus* (Peters). *Bulletin of Environmental Contamination and Toxicology*, **36**, 888–94.

Sellers, C.M., Heath, A.G. & Bass, M.L. (1975). The effect of sublethal concentrations of copper and zinc on ventilatory activity, blood oxygen and pH in rainbow trout (*Salmo gairdneri*). *Water Research*, **1**, 401–8.

Settle, D.M. & Patterson, C.C. (1980). Lead in Albacore: guide to lead pollution in Americans. *Science (New York)*, **207**, 1167–76.

Sharpe, V. & Denny, P. (1976). Electron microscope studies on the absorption and localization of lead in the leaf tissue of *Potamogeton pectinatus* L. *Journal of Experimental Botany*, **27**, 1155–62.

Shaw, P.A. & Maga, J.A. (1943). The effect of mining silt on yield of fry from salmon spawning beds. *California Fish & Game*, **29**, 29–41.

Shaw, T.L. & Brown, V.M. (1974). The toxicity of some forms of copper to rainbow trout. *Water Research*, **8**, 377–82.

Sheath, R.G., Havas, M., Hellebust, J.A. & Hutchinson, T.C. (1982). Effect of long-term acidification on the algal communities of tundra ponds at the Smoking Hills, N.W.T., Canada. *Canadian Journal of Botany,* **60**, 58–72.

Shehata, F.H.A. & Whitton, B.A. (1982). Zinc tolerance in strains of the blue-green alga *Anacystis nidulans*. *British Phycological Journal,* **17**, 5–12.

Shiller, A.M. & Boyle, E. (1985). Dissolved zinc in rivers. *Nature (London),* **317**, 49–52.

Sibley, T.H. & Morgan, J.J. (1975). Equilibrium speciation of trace metals in freshwater: seawater mixtures. pp. 319–38. In: Hutchinson, T.C. (Ed.) *Symposium Proceedings of the International Conference on Heavy Metals in the Environment Volume 1: Plenary Futures, Analytical.* Toronto, Ontario, Canada.

Sigg, L. (1985). Metal transfer mechanisms in lakes; the role of settling particles. pp. 283–310. In: Stumm, W. (Ed.) *Chemical Process in Lakes.* Wiley-Interscience, New York.

Silverberg, B.A. (1975). Ultrastructural localization of lead in *Stigeoclonium tenue* (Chlorophyceae, Ulotrichales) as demonstrated by cytochemical and X-ray microanalysis. *Phycologia,* **14**, 265–74.

Silverberg, B.A. (1976). Cadmium-induced ultrastructural changes in mitochondria of freshwater green algae. *Phycologia,* **15**, 155–9.

Singer, P.C. & Stumm, W. (1970). Acidic mine drainage: the rate-determining step. *Science (New York),***167**, 1121–3.

Singh, S.P. & Pandey, A.K. (1982). Cadmium-mediated resistance to metals and antibiotics in a cyanobacterium. *Molecular and General Genetics,* **187**, 240–3.

Singh, S.P. & Yadava,, V. (1983). Cadmium induced inhibition of nitrate uptake in *Anacystis nidulans*: interaction with other divalent cations. *Journal of General and Applied Microbiology,* **29**, 297–304.

Singh, S.P & Yadava, V. (1984). Cadmium induced inhibition of ammonium and phosphate uptake in *Anacystis nidulans*: interaction with other divalent cations. *Journal of General and Applied Microbiology,* **30**, 79–86.

Singh, S.P. & Yadava, V. (1986). Cadmium tolerance in the cyanobacterium *Anacystis nidulans. Biologisches Zentralblatt,* **105**, 539–42.

Skelly and Loy (Consultants) & Pennsylvania Environmental Consultants (1973). *Processes, Procedures and Methods to Control Pollution from Mining Activities: Surface Mining.* 390 pp. EPA Publication No. EPA-430/9-73-011, Washington.

Skowronski, T. (1984). Energy dependent transport of cadmium by *Stichococcus bacillaris. Chemosphere,* **13**, 1379–84.

Smith, O.R. (1940). Placer mining silt and its relation to salmon and trout on the Pacific coast. *Transactions of the American Fisheries Society,* **69**, 225–30.

Smock, L.A. (1983*a*). The influence of feeding habits on whole-body metal concentrations in aquatic insects. *Freshwater Biology,* **13**, 301–11.

Smock, L.A. (1983*b*). Relationships between metal concentrations and organism size in aquatic insects. *Freshwater Biology,* **13**, 313–21.

Solbé, J.F. de L.G. (1974). The toxicity of zinc sulphate to rainbow trout in very hard water. *Water Research,* **8**, 389–91.

Solbé, J.F. de L.G. (1977). Water quality, fish and invertebrates in a Zn-polluted stream. pp. 97–105. In: Alabaster, J.S. (Ed.) *Biological Monitoring of Inland Fisheries.* Applied Science Publishers, London.

Solbé, J.F. de L.G. & Cooper, V.A. (1976). Studies on the toxicity of copper sulphate to stone loach *Noemacheilus barbatulus* (L.) in hard water. *Water Research*, **10**, 523–7.

Solbé, J.F. de L.G. & Flook, V.A. (1975). Studies on the toxicity of zinc sulphate and of calcium sulphate to stone loach *Noemacheilus barbatulus* (L.) in hard water. *Journal of Fish Biology*, **7**, 631–7.

Sorentino, C. (1985). Copper resistance in *Hormidium fluitans* (Gay) Heering (Ulotrichaceae, Chlorophyceae). *Phycologia*, **24**, 366–8.

Sparks, R.E., Waller, W.T. & Cairns, J. (1972). Effect of shelters on the resistance of dominant and submissive bluegills (*Lepomis macrochirus*) to a lethal concentration of zinc. *Journal of the Fisheries Research Board of Canada*, **29**, 1356–8.

Spehar, R.L., Anderson, R.L. & Fiandt, J.T. (1978). Toxicity and bioaccumulation of cadmium and lead in aquatic invertebrates. *Environmental Pollution*, **15**, 195–208.

Spencer, D.F. & Nichols, L.H. (1983). Free nickel ion inhibits growth of two species of green algae. *Environmental Pollution Series A*, **31**, 97–104.

Speranza, A.W., Seeley, R.J., Seeley, V.A. & Perlmutter, A. (1977). The effect of zinc on reproduction in the zebrafish *Brachydanio rerio* Hamilton-Buchanan. *Environmental Pollution*, **12**, 217–22.

Sprague, J.B., Elson, P.F. & Saunders, R.L. (1965). Sublethal copper–zinc pollution in a salmon river — a field and laboratory study. *International Journal of Air and Water Pollution*, **9**, 531–43.

Starodub, M.E., Wong, P.T.S., Mayfield, C.I. & Chau, Y.K. (1987). Influence of complexation and pH on individual and combined heavy metal toxicity to a freshwater green alga. *Canadian Journal of Fisheries and Aquatic Science*, **44**, 1173–80.

Stauber, J.L. & Florence, T.M. (1987). Mechanism of toxicity of ionic copper and copper complexes to algae. *Marine Biology*, **94**, 511–19.

Steemann-Nielson, E. & Wium-Anderson, S. (1970). Copper ions as poisons in the sea and freshwater. *Marine Biology*, **6**, 93–7.

Steinback, J.T. (1966). An ecological investigation of the algal genera and other biota in waters polluted by mineral acid-drainage from mines in Vintan Co., Ohio. MSc thesis, Ohio State University.

Stiff, M.J. (1971). The chemical states of copper in polluted freshwater and a scheme of analysis to differentiate them. *Water Research*, **5**, 585–99.

Stockinger, N.F. & Hays, H.A. (1960). Plankton, benthos and fish in three strip-mine lakes with varying pH values. *Transactions of the Kansas Academy of Science*, **63**, 1–11.

Stokes, P.M. (1975*a*). Adaptation of green algae to high levels of copper and nickel in aquatic environments. pp. 137–54. In: Hutchinson, T.C. (Ed.) *Symposium Proceedings of the International Conference on Heavy Metals in the Environment Volume II part I: Pathways and Cycling*. Toronto, Ontario, Canada, October 27–31, 1975.

Stokes, P.M. (1975*b*). Uptake and accumulation of copper and nickel by metal-tolerant strains of *Scenedesmus*. *Verhandlungen, Internationale Vereinigung für Theoretische und Angewandte Limnologie*, **19**, 2128–37.

Stokes, P.M. (1979). Copper accumulations in freshwater biota. pp. 360–81. In: Nriagu, J.O. (Ed.) *Copper in the Environment Part 1: Ecological Cycling*. Wiley-Interscience, New York.

Stokes, P.M. (1981). Nickel in aquatic systems. In: *Effects of Nickel in the Canadian Environment.* National Research Council, Canada Report No. 18568.

Stokes, P.M. (1983). Responses of freshwater algae to metals. pp. 87–112. In: Round, F.E. & Chapman, D.J. (Eds) *Progress in Phycological Research Volume 2.* Elsevier, Amsterdam.

Stokes, P.M., Hutchinson, T.C. & Krauter, K. (1973). Heavy-metal tolerance in algae isolated from contaminated lakes near Sudbury, Ontario. *Canadian Journal of Botany,* **51**, 2155–68.

Stumm, W. & Bilinski, H. (1973). Trace metals in natural waters: difficulties of interpretation arising from our ignorance of their speciation. pp. 39–49. In: Jenkins, S.H. (Ed.) *Advances in Water Pollution Research.* Sixth International Conference held in Jerusalem, June 8–23, 1972, Pergamon Press, Oxford & New York.

Stumm, W. & Morgan, J.J. (1981). Aquatic Chemistry. *An Introduction Emphasizing Chemical Equilibria in Natural Waters.* 2nd Edition. 780 pp. Wiley-Interscience, New York.

Swain, R. & White, R.W.G. (1985). Influence of a metal-contaminated tributary on the invertebrate drift fauna of the King River (Tasmania, Australia). *Hydrobiologia,* **122**, 261–6.

Sylva, R.N. (1976). The environmental chemistry of copper(II) in aquatic systems. *Water Research,* **10**, 789–92.

Szeto, C. & Nyberg, D. (1979). The effect of temperature on copper tolerance of *Paramaecium. Bulletin of Environmental Contamination and Toxicology,* **21**, 131–5.

Tanaka, O., Nasu, Y., Yanase, D., Takimoto, A. & Kugimoto, M. (1982). pH dependence of the copper effect on flowering, growth and chlorophyll content in *Lemna paucicostata* 6746. *Plant and Cell Physiology,* **23**, 1479–82.

Taylor, B.E., Wheeler, M.E. & Nordstrom, D.K. (1984*a*). Stable isotope geochemistry of acid mine drainage: experimental oxidation of pyrite. *Geochimica et Cosmochimica Acta,* **48**, 2669–78.

Taylor, B.E., Wheeler, M.E. & Nordstrom, D.K. (1984*b*). Isotope composition of sulphate in acid mine drainage as a measure of bacterial oxidation. *Nature (London),* **308**, 538–41.

Taylor, G.J. & Crowder, A.A. (1984). Copper and nickel tolerance in *Typha latifolia* clones from contaminated and uncontaminated ecosystems. *Canadian Journal of Botany,* **62**, 1304–8.

Tessier, A. & Campbell, P.G.C. (1987). Partitioning of trace metals in sediments: relationships with bioavailability. *Hydrobiologia,* **149**, 43–52.

Tewari, H., Gill, T.S. & Pant, J. (1987) Impact of chronic lead poisoning on the haematological and biochemical profiles of a fish, *Barbus conchonius. Bulletin of Environmental Contamination and Toxicology,* **38**, 748–52.

Thompson, K.W., Hendricks, A.C. & Cairns, J. (1980). Acute toxicity of zinc and copper singly and in combination to the Bluegill (*Lepomis macrochirius. Bulletin of Environmental Contamination and Toxicology,* **25**, 122–9.

Thornton, I. (1975). Applied geochemistry in relation to mining and the environment. pp. 87–102. In: Jones, M.T. (Ed.) *Minerals and the Environment.* Institution of Mining and Metallurgy, London.

Thorp, V.J. & Lake, P.S. (1973). Pollution of a Tasmanian river by mine effluents. II. Distribution of macroinvertebrates. *Internationale Revue der Gesamten Hydrobiologie,* **58**, 885–92.

Thorp, V.J. & Lake, P.S. (1974). Toxicity bioassays of cadmium on selected freshwater invertebrates and the interaction of cadmium and zinc on the freshwater shrimp, *Paratya tasmaniensis* Riek. *Australian Journal of Marine and Freshwater Research,* **25**, 97–104.

Tipping, E., Griffith, J.R. & Hilton, J. (1983). The effect of adsorbed humic substances on the uptake of copper(II) by goethite. *Croatica Chemica Acta,* **56**, 613–21.

Tipping, E., Hetherington, N.B., Hilton, J., Thompson, D.W., Bowles, E. & Hamilton-Taylor, J. (1985). Artifacts in the use of selective chemical extraction to determine distributions of metals between oxides of manganese and iron. *Analytical Chemistry,* **57**, 1944–6.

Tipping, E., Thompson, D.W., Ohnstad, M. & Hetherington, N.B. (1986). Effects of pH on the release of metals from naturally-occurring oxides of Mn and Fe. *Environmental Technology Letters,* **7**, 109–14.

Tomkiewicz, S.M. & Dunson, W.A. (1977). Aquatic insect diversity and biomass in a stream marginally polluted by acid strip mine drainage. *Water Research,* **11**, 397–402.

Trefry, J.H., Metz, S., Trocine, R.P. & Nelsen, T.A. (1985). A decline in lead transported by the Mississippi River. *Science (New York),* **230**, 439–41.

Trollope, D.R. & Evans, B. (1976). Concentrations of Cu, Fe, Pb, Ni and Zn in freshwater algal blooms. *Environmental Pollution,* **11**, 109–16.

Tsai, C.-F. (1979). Survival, overturning and lethal exposure times for the pearl dace, *Semotilus margaritus* (Cope), exposed to copper solution. *Comparative Biochemistry and Physiology*, **64C**, 1–6.

Turekian, K.K. (1971). Rivers, tributaries, and estuaries. pp. 9–74. In: Hood, D.W. (Ed.) *Impingement of Man on the Oceans.* Wiley-Interscience, New York.

Turekian, K.K. & Scott, M.R. (1967). Concentrations of chromium, silver, molybdenum, nickel, cobalt and manganese in suspended material in streams. *Environmental Science and Technology,* **1**, 940–2.

Tyler, P.A. & Buckney, R.T. (1973). Pollution of a Tasmanian river by mine effluents. I. Chemical evidence. *Internationale Revue der Gesamten Hydrobiologie,* **58**, 873–83.

Van Duyn-Henderson, J.A. & Lasenby, D.C. (1986). Zinc and cadmium transport by the vertically migrating Opossum Shrimp, *Mysis relicta. Canadian Journal of Fisheries and Aquatic Science,* **43**, 1726–32.

Van Everdingen, R.O. & Krouse, H.R. (1985). Isotope composition of sulphates generated by bacterial and abiological oxidation. *Nature (London),* **315**, 395–6.

Van Nieuwenhuyse, E.E. & LaPerriere, J.D. (1986). Effects of placer gold mining on primary production in subarctic streams in Alaska. *Water Resources Bulletin,* **22**, 91–9.

Viale, G. & Calamari, D. (1984). Immune response in rainbow trout *Salmo gairdneri* after long-term treatment with low levels of Cr, Cd and Cu. *Environmental Pollution Series A,* **35**, 247–57.

Vinikour, W.S. (1979). Coal slurry observed as a habitat for semi-aquatic beetle

Laternarius brunneus (Coleoptera, Heteroceridae) with notes on water quality conditions. *Entomological News*, **90**, 203–4.

Vuceta, J. & Morgan, J.J. (1978). Chemical modeling of trace metals in fresh waters: role of complexation and adsorption. *Environmental Science and Technology*, **12**, 1302–9.

Wagemann, R. & Barica, J. (1979). Speciation and rate of loss of copper from lakewater with implications to toxicity. *Water Research*, **13**, 515–23.

Wakao, N., Tachibana, H., Tanaka, Y., Sakurai, Y. & Shiota, H. (1985). Morphological and physiological characteristics of streamers in acid mine drainage waters from a pyritic mine. *Journal of General and Applied Microbiology*, **31**, 17–28.

Wallwork, J.F. & Hunter, R.W. (1981). Discussion contribution to paper 'Heavy metals in the Derwent Reservoir catchment'. pp. 87–91. In: Say, P.J. & Whitton, B.A. (Eds) *Heavy Metals in Northern England: Environmental and Biological Aspects*. Department of Botany, University of Durham, England.

Wang, H.K. & Wood, J.M. (1984). Bioaccumulation of nickel by algae. *Environmental Science and Technology*, **18**, 106–9.

Warner, R.W. (1968). Preliminary report on the biology of acid mine drainage at Grassy Run and Roaring Creek, W. Virginia. Federal Water Pollution Control Administration, Cincinnati, Ohio, USA.

Warner, R.W. (1971). Distribution of biota in a stream polluted by acid mine drainage. *Ohio Journal of Science*, **71**, 202–15.

Warnick, S.L. & Bell, H.L. (1969). The acute toxicity of some heavy metals to different species of aquatic insect. *Journal of the Water Pollution Control Federation*, **41**, 280–4.

Watkin, E.M. (1983). Revegetation and water quality aspects of mine and tailings drainage control. 21 pp. Typescript of paper prepared for presentation to the annual symposium, West Virginia Acid Mine Drainage Task Force, Clarksburg, West Virginia, 26/5/1983.

Watson, R.B. (1983). Rapid drainage of coal seams for prevention of acid formation. pp. 103–14. In: Odendaal, P.E. (Ed.) *Mine Water Pollution Proceeds — International Association of Water Pollution Research*. Water Science and Technology (South Africa)**15**(2), 1–181.

Watton, A.J. & Hawkes, H.A. (1984). The acute toxicity of ammonia and copper to the gastropod *Potamopyrgus jenkinsi* (Smith).*Environmental Pollution Series A*, **36**, 17–29.

Weatherley, A.H. Beevers, J.R. & Lake, P.S. (1967). The ecology of a zinc-polluted river. pp. 252–78. In: Weatherley A.H. (Ed.) *Australian Inland Waters and their Faunas: Eleven Studies*. Australian National University Press, Canberra.

Weatherley, A.H., Lake, P.S. & Rogers, S.C. (1980). Zinc pollution and the ecology of the freshwater environment. pp. 337–418. In: Nriagu, J.O. (Ed.) *Zinc in the Environment Part 1: Ecological Cycling*. Wiley-Interscience, New York.

Weed, C.E. & Rutschky, C.W. (1972). Benthic macroinvertebrate community structure in a stream receiving acid mine drainage. *Proceedings of the Pennsylvania Academy of Science*, **46**, 41–7.

Wehr, J.D. & Whitton, B.A. (1983*a*). Aquatic cryptograms of acid natural springs enriched with heavy metals: The Kootenay Paint Pots, British Columbia. *Hydrobiologia*, **98**, 97–105.

Wehr, J.D. & Whitton, B.A. (1983*b*). Accumulation of heavy metals by aquatic mosses. 2. *Rhynchostegium riparioides*. *Hydrobiologia*, **100**, 261–84.

Wehr, J.D., Kelly, M.G. & Whitton, B.A. (1987). Factors affecting accumulation and loss of zinc by the aquatic moss *Rhynchostegium riparioides*. *Aquatic Botany*, **29**, 261–74.

Welsh, R.P.H. & Denny, P. (1976). Waterplants and the recycling of heavy metals in an English Lake. pp. 217–23. In Hemphill, D.D. (Ed.) *Trace Substances in Environmental Health — X*. University of Missouri, Columbia.

Welsh, R.P.H. & Denny, P. (1979). The translocation of lead and copper in two submerged aquatic angiosperm species. *Journal of Experimental Botany*, **30**, 339–45.

Welsh, R.P.H. & Denny, P. (1980). The uptake of lead and copper by submerged aquatic macrophytes in two English lakes. *Journal of Ecology*, **68**, 443–55.

Wentsel, R., McIntosh, A. & McCafferty, W.P. (1978*a*). Emergence of the midge *Chironomus tentans* when exposed to heavy metal contaminated sediment. *Hydrobiologia*, **57**, 195–6.

Wentsel, R., McIntosh, A. & Atchison, G. (1978*b*). Evidence of resistance to metals in larvae of the midge *Chironomus tentans* in a metal contaminated lake. *Bulletin of Environmental Contamination and Toxicology*, **20**, 451–5.

Westall, J.C., Zachary, J.L. & Morel, F.M.M. (1976). *MINEQL, a computer program for the calculation of the chemical equilibrium composition of aqueous systems*. 91 pp. MIT Department of Civil Engineering Technical Report No. 18.

Whiteley, L.S. (1968). The resistance of tubificial worms to three common pollutants. *Hydrobiologia*, **32**, 193–205.

Whitton, B.A. (1970). Toxicity of zinc, copper and lead to Chlorophyta from flowing waters. *Archiv für Mikrobiologie*, **72**, 353–60.

Whitton, B.A. (1980). Zinc and plants in rivers and streams. pp. 363–400. In: Nriagu, J.O. (Ed.) *Zinc in the Environment Part II: Health Effects*. Wiley-Interscience, New York.

Whitton, B.A. (1984). Algae as monitors of heavy metals. pp. 257–80. In: Shubert, L.E. (Ed.). *Algae as Ecological Indicators*. Academic Press, London.

Whitton, B.A. & Diaz, B.M. (1980). Chemistry and plants of streams and rivers with elevated Zn. pp. 457–63. In: Hemphill, D.D. (Ed.) *Trace Substances in Environmental Health — XIV*. University of Missouri, Columbia.

Whitton, B.A. & Diaz, B.M. (1981). Influence of environmental factors on photosynthetic species composition in highly acidic waters. *Verhandlungen, Internationale Vereinigung für Theoretische und Angewandte Limnologie*, **21**, 1459–65.

Whitton, B.A. & Shehata, F.H.A. (1982). Influence of cobalt, nickel, copper and cadmium on the blue-green alga *Anacystis nidulans*. *Environmental Pollution Series A*, **27**, 275–81.

Whitton, B.A., Gale, N.L. & Wixson, B.G. (1981*a*). Chemistry and plant ecology of zinc-rich wastes dominated by blue-green algae.*Hydrobiologia*, **83**, 331–41.

Whitton, B.A., Say, P.J. & Wehr, J.D. (1981*b*). Use of plants to monitor heavy metals in rivers. pp. 135–46. In: Say, P.J. & Whitton, B.A. (Eds) *Heavy Metals in Northern England: Environmental and Biological Aspects*. Department of Botany, University of Durham, England.

Whitton, B.A., Say, P.J. & Jupp, B.P. (1982). Accumulation of zinc, cadmium and lead by the aquatic liverwort *Scapania*. *Environmental Pollution Series B*, **3**, 299–316.

Wickham, P., van de Walle, E. & Planas, D. (1987). Comparative effects of mine wastes on the benthos of an acid and an alkaline pool. *Environmental Pollution*, **44**, 83–9.

Wilcox, G. & DeCosta, J. (1982). The effect of phosphorus and nitrogen additions on the algal biomass and species composition of an acidic lake. *Archiv für Hydrobiologie*, **94**, 393–424.

Wilcox, G.R. & DeCosta, J. (1984). Bag experiments on the effect of phosphorus and base additions on the algal biomass and species composition of an acid lake. *International Revue der Gesamten Hydrobiologie*, **69**, 173–99.

Wilkins, P. (1977). Observations on ecology of *Mielichhoferia elongata* and other 'copper mosses' in the British Isles. *Bryologist*, **80**, 175–81.

Williams, K.A., Green, D.W.J., Pascoe, D. & Gower, D.E. (1986). The acute toxicity of cadmium to different larval stages of *Chironomus riparius* (Diptera, Chironomidae) and its ecological significance for pollution regulation. *Oecologia*, **70**, 362–6.

Williams, L.G. & Mount, D.I. (1965). Influence of zinc on periphytic communitites. *American Journal of Botany*, **52**, 26–34.

Willis, M. (1985). A comparative survey of the *Erpobdella octoculata* (L.) populations in the Afon Crafnant, N. Wales, above and below an input of zinc from mine waste. *Hydrobiologia*, **120**, 107–18.

Wilson, A.L. (1976). *Concentrations of trace metals in river waters: a review*. Water Research Centre Technical Report No. TR 16. Medmenham.

Wilson, R.C.H. (1972). Prediction of copper toxicity in receiving waters. *Journal of the Fisheries Research Board of Canada*, **29**, 1500–2.

Winner, R.W. (1984). The toxicity and bioaccumulation of cadmium and copper as affected by humic acid. *Aquatic Toxicology*, **5**, 267–74.

Winner, R.W. & Gauss, J.D. (1986). Relationship between chronic toxicity and bioaccumulation of copper, cadmium and zinc as affected by water hardness and humic acid. *Aquatic Toxicology*, **8**, 149–61.

Winner, R.W., Kelling, T., Yeager, R. & Farrell, M.P. (1977). Effects of food type on the acute and chronic toxicity of copper to *Daphnia magna*. *Freshwater Biology*, **7**, 343–9.

Wixson, B.G. (1972). *The Missouri Lead Study. An interdisciplinary investigation of environmental pollution by lead and other heavy metals by industrial development in the New Lead Belt of Southeastern Missouri*. 2 Volumes. 1108 pp. Report submitted to NSF-RANN by the University of Missouri-Rolla.

Wolfe, J.A. (1985). *Mineral Resources. A World Review*. 293 pp. Chapman & Hall, New York and London.

Woltering, D.M. (1984). The growth response in fish chronic and early life stages toxicity tests: a critical review. *Aquatic Toxicology*, **5**, 1–21.

Wong, P.T.S., Chau, Y.K. & Luxon, P.L. (1975). Methylation of lead in the environment. *Nature (London)*, **253**, 263–4.

Wong, P.T.S., Chau, Y.K., Yaromich, J.L. & Kramar, O. (1987). Bioaccumulation and methylation of tri- and dialkyllead compounds by a freshwater alga. *Canadian Journal of Fisheries and Aquatic Science*, **44**, 1257–60.

Wong, S.L. & Beaver, J.L. (1980). Algal bioassays to determine toxicity of metal mixtures. *Hydrobiologia*, **74**, 199–208.
Wong, S.L. & Beaver, J.L. (1981). Metal interactions in algal toxicology, conventional versus *in vivo* tests. *Hydrobiologia*, **85**, 65–71.
Wood, J.M. & Wang, H.K. (1985). Strategies for microbial resistance to heavy metals. pp. 81–98. In: Stumm, W. (Ed.) *Chemical Processes in Lakes*. Wiley-Interscience, New York.
Wright, D.A. (1980). Cadmium and calcium interactions in the freshwater amphipod *Gammarus pulex*. *Freshwater Biology*, **10**, 123–33.
Wurtsbaugh, W.A. & Home, A.J. (1982). Effects of copper on nitrogen fixation and growth of blue-green algae in natural plankton associations. *Canadian Journal of Fisheries and Aquatic Science*, **39**, 1636–41.
Wurtz, C.B. (1962). Zinc effects in freshwater mollusks. *Nautilus*, **76**, 53–61.
Wurtz, C.B. & Bridges, C.H. (1961). Preliminary results from macroinvertebrate bioassays. *Proceedings of the Pennyslvania Academy of Sciences*, **35**, 51–6.
Yager, C.M. & Harry, H.W. (1964). The uptake of radioactive zinc, cadmium and copper by the freshwater snail, *Taphius glabratus*. *Malecologia*, **1**, 339–53.
Yan, N.D., Miller, G.E., Wile, I. & Hitchin, G.G. (1985). Richness of aquatic macrophyte floras of soft water lakes of differing pH and trace metal content in Ontario, Canada. *Aquatic Botany*, **23**, 27–40.
Yasuno, M., Hatakeyama, S. & Sugaya, Y. (1985). Characteristic distribution of chironomids in the rivers polluted with heavy metals. *Verhandlungen, Internationale Vereinigung für Theoretische und Angewandte Limnologie*, **22**, 2371–7.
Zitko, V. & Carson, W.G. (1976). A mechanism of the effects of water hardness on the lethality of heavy metals to fish. *Chemosphere*, **5**, 299–303.

Index